Geometry

LARSON
BOSWELL
STIFF

Applying • Reasoning • Measuring

UNIVERSITY OF SAINT FRANCIS
DEPARTMENT OF EDUCATION
WITHDRAWN

Practice Workbook with Examples Teacher's Edition

The Practice Workbook provides additional practice with worked-out examples for every lesson. The workbook covers essential skills and vocabulary. Space is provided for students to show their work. The Teacher's Edition includes the student workbook and the answers.

McDougal Littell
A HOUGHTON MIFFLIN COMPANY
Evanston, Illinois • Boston • Dallas

Copyright © 2001 McDougal Littell Inc.
All rights reserved.

No part of this work may be reproduced or transmitted in any form
or by any means, electronic or mechanical, including photocopying
and recording, or by any information storage or retrieval system,
without the prior written permission of McDougal Littell Inc. unless
such copying is expressly permitted by federal copyright law.
Address inquiries to Manager, Rights and Permissions, McDougal
Littell Inc., P.O. Box 1667, Evanston, IL 60204.

ISBN: 0-618-04332-2

456789-VEI- 04 03 02

Contents

Chapter	Title	
1	Basics of Geometry	1-21
2	Reasoning and Proof	22-39
3	Perpendicular and Parallel Lines	40-60
4	Congruent Triangles	61-81
5	Properties of Triangles	82-99
6	Quadrilaterals	100-120
7	Transformations	121-138
8	Similarity	139-159
9	Right Triangles and Trigonometry	160-180
10	Circles	181-201
11	Area of Polygons and Circles	202-219
12	Surface Area and Volume	220-240
	Answers	A1-A11

LESSON 1.1

NAME _____ DATE _____

Practice with Examples

For use with pages 3–9

GOAL Find and describe patterns and use inductive reasoning

VOCABULARY

A **conjecture** is an unproven statement that is based on observations.

Inductive reasoning is a process that involves looking for patterns and making conjectures.

A **counterexample** is an example that shows a conjecture is false.

EXAMPLE 1 Describing a Visual Pattern

Sketch the next figure in the pattern.

 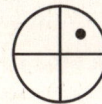 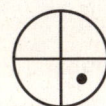

SOLUTION

Each figure looks like the one before it except that it has rotated 90°. The next figure will have the smaller circle in the lower-left quarter of the bigger circle.

Exercise for Example 1

1. Sketch the next figure in the pattern.

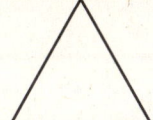 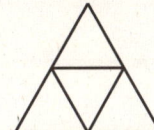 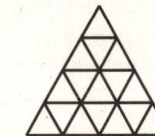

Copyright © McDougal Littell Inc.
All rights reserved.

Geometry
Practice Workbook with Examples

LESSON 1.1 CONTINUED

Practice with Examples
For use with pages 3–9

EXAMPLE 2 Describing a Number Pattern

Describe a pattern in the sequence of numbers. Predict the next number.

a. $5, 3, 1, -1, \ldots$ **b.** $1, -4, 9, -16, \ldots$ **c.** $\frac{1}{2}, \frac{1}{4}, \frac{1}{8}, \ldots$

SOLUTION

a. These are consecutive odd numbers, but listed backwards starting with 5. The next number is -3.

b. These numbers look like consecutive perfect squares, except that every other one is negative. The next number is 25.

c. Each number is $\frac{1}{2}$ times the previous number. The next number is $\frac{1}{16}$.

Exercises for Example 2

Describe a pattern in the sequence of numbers. Predict the next number.

2. $1, 2, 6, 24, \ldots$ **3.** $0, 3, 8, 15, 24, \ldots$

EXAMPLE 3 Making a Conjecture

Complete the conjecture.

Conjecture: The product of two consecutive even integers is divisible by ___?___.

SOLUTION

List some specific examples and look for a pattern.

Examples:

$2 \times 4 = 8 = 8 \times 1$ $6 \times 8 = 48 = 8 \times 6$ $10 \times 12 = 120 = 8 \times 15$

$4 \times 6 = 24 = 8 \times 3$ $8 \times 10 = 80 = 8 \times 10$ $12 \times 14 = 168 = 8 \times 21$

Conjecture: The product of two consecutive even integers is divisible by 8.

LESSON 1.1 CONTINUED

Practice with Examples

For use with pages 3–9

Exercises for Example 3

Complete the conjecture based on the pattern you observe in the specific cases.

4. **Conjecture:** For any two numbers a and b, the product of $(a + b)$ and $(a - b)$ is always equal to ___?___.

   $(2 + 1) \times (2 - 1) = 3 = 2^2 - 1^2$ $(4 + 2) \times (4 - 2) = 12 = 4^2 - 2^2$
   $(3 + 2) \times (3 - 2) = 5 = 3^2 - 2^2$ $(6 + 3) \times (6 - 3) = 27 = 6^2 - 3^2$

EXAMPLE 4   Finding a Counterexample

Show the conjecture is false by finding a counterexample.

Conjecture: All odd numbers are prime.

SOLUTION

The conjecture is false. Here is a counterexample: The number 9 is odd and is a composite number, not a prime number.

Exercise for Example 4

Show the conjecture is false by finding a counterexample.

5. The square of the sum of two numbers is equal to the sum of the squares of the two numbers. That is, $(a + b)^2 = a^2 + b^2$.

LESSON 1.2

Practice with Examples
For use with pages 10–16

GOAL Understand and use the basic undefined terms and defined terms of geometry and sketch intersections of lines and planes

VOCABULARY

A **point** has no dimension, a **line** extends in one dimension, and a **plane** extends in two dimensions.

Collinear points are points that lie on the same line.

Coplanar points are points that lie on the same plane.

On a line passing through points A and B, **segment** AB consists of all points between A and B and **endpoints** A and B.

On a line passing through points A and B, **ray** AB consists of the **initial point** A and all points on the same side of A as point B.

If point C is between A and B, then ray CA and ray CB are **opposite rays.**

Two or more geometric figures **intersect** if they have one or more points in common. The **intersection** of the figures is the set of points the figures have in common.

EXAMPLE 1 Drawing and Naming Lines, Segments, and Rays

a. Draw three noncollinear points, A, B, and C. Then draw point D on line AB between points A and B. Draw segment CD. Draw ray CA and ray CB.
b. Are points A, B, and D collinear? Are points B, C, and D collinear?
c. Are ray CA and ray CB opposite rays? Are ray DA and ray DB opposite rays?

SOLUTION

a.

[Four diagrams showing progression: A, B, C as separate points; then line AB drawn; then D added on line AB; with C off the line]

1. First, draw A, B, and C. 2. Draw line AB. 3. Draw D.

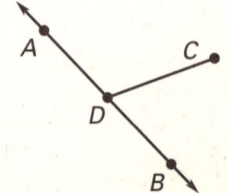

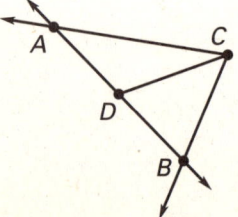

4. Draw segment CD. 5. Draw ray CA and ray CB.

LESSON 1.2 CONTINUED

Practice with Examples
For use with pages 10–16

b. Yes, points *A*, *B*, and *D* are collinear because they lie on line *AB*. No, points *B*, *C*, and *D* are noncollinear because a straight line cannot be drawn through all three points.

c. No, ray *CA* and ray *CB* are not opposite rays. Point *C* is not between *A* and *B*. Yes, ray *DA* and ray *DB* are opposite rays. Point *D* is between *A* and *B*.

Exercise for Example 1

1. Draw collinear points *A*, *B*, and *C*, with point *B* between *A* and *C*. Draw point *D* not on line *AC*. Draw line *AD*. Draw point *E* on line *AD* between point *A* and point *D*. Draw segment *EC*. Draw ray *BD*. Draw ray *EB*.

Use the diagram to name the figures.

2. Three noncollinear points

3. Two opposite rays

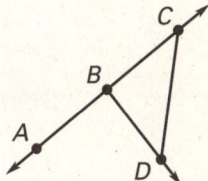

4. One line segment

5. Three collinear points

6. Two rays which are *not* opposite rays

7. Two line segments that are on the same line

LESSON 1.2 CONTINUED

Practice with Examples
For use with pages 10–16

EXAMPLE 2 Sketching Intersections

Sketch the figure described.

a. Three lines that lie in the same plane, but two of the lines do not intersect with each other and the third line intersects with each of the other lines in a point.

b. Two planes which do not intersect, and a line which intersects each plane in a point.

SOLUTION

a.

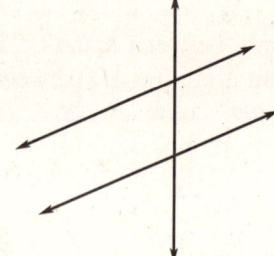

Draw two lines which do not intersect. Draw a third line, crossing each of the other lines.

b.

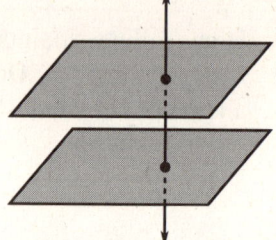

Draw two planes which do not intersect. Draw a line through both planes. Emphasize the points where the line intersects.

Exercises for Example 2

Sketch the figure described.

8. Three planes which intersect in a line

9. Two planes which intersect in a line, and a third plane which intersects each of the other two planes in a line, but not the same line

LESSON 1.3

NAME _____ DATE _____

Practice with Examples
For use with pages 17–24

GOAL Use segment postulates and use the distance formula to measure distances

VOCABULARY

A **postulate** or **axiom** is a rule that is accepted without proof.

Postulate 1 Ruler Postulate:
The points on a line can be matched one to one with the real numbers. The real number that corresponds to a point is the **coordinate** of the point.

The **distance** between points A and B, written as AB, is the absolute value of the difference between the coordinates of A and B.

AB is also called the **length** of $\overline{AB}$.

When three points lie on a line, you can say that one of them is **between** the other two.

Postulate 2 Segment Addition Postulate:
If B is between A and C, then $AB + BC = AC$. If $AB + BC = AC$, then B is between A and C.

The **Distance Formula** is a formula for computing the difference between two points in a coordinate plane.

The Distance Formula:
If $A(x_1, y_1)$ and $B(x_2, y_2)$ are points in a coordinate plane, then the distance between A and B is $AB = \sqrt{(x_2 - x_1)^2 + (y_2 - y_1)^2}$.

Segments that have the same length are called **congruent segments**.

EXAMPLE 1 Using the Segment Addition Postulate

In the diagram of the collinear points, $DE = 2$, $EF = 3$, and $DE = FG$. Find each length.
FG
DF
DG
EG

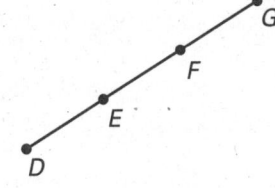

SOLUTION

Since $DE = FG$ and $DE = 2$, $FG = 2$.

Since $DF = DE + EF$, $DF = 2 + 3 = 5$.

Since $DG = DF + FG$, $DG = 5 + 2 = 7$.

Since $EG = EF + FG$, $EG = 3 + 2 = 5$.

LESSON 1.3 CONTINUED

Practice with Examples

For use with pages 17–24

Exercises for Example 1

1. In the diagram of the collinear points, $BC = 5$ and $BC = AB$. Find the following lengths.
   a. AC
   b. AB
   c. Are any segments congruent?

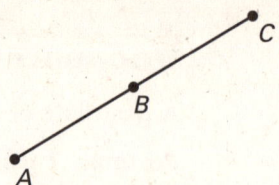

2. In the diagram of the collinear points, $HK = 9$, $HI = JK$, and $IJ = 1$. Find the following lengths.
   a. HI
   b. JK
   c. HJ
   d. IK

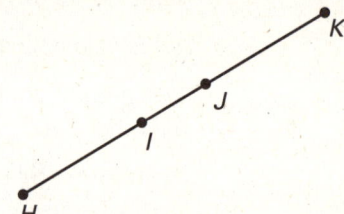

EXAMPLE 2 Using the Distance Formula

Find the following distances. State whether any of the segments are congruent.
   a. AB
   b. BC
   c. CD
   d. AC

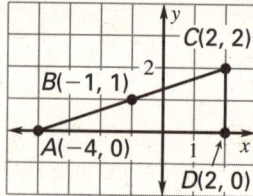

SOLUTION

Use the Distance Formula.

a. $AB = \sqrt{[(-1) - (-4)]^2 + (1 - 0)^2} = \sqrt{3^2 + 1^2} = \sqrt{9 + 1} = \sqrt{10}$

b. $BC = \sqrt{[2 - (-1)]^2 + (2 - 1)^2} = \sqrt{3^2 + 1^2} = \sqrt{9 + 1} = \sqrt{10}$

c. $CD = \sqrt{(2 - 2)^2 + (0 - 2)^2} = \sqrt{0^2 + (-2)^2} = \sqrt{0 + 4} = \sqrt{4} = 2$

d. $AC = \sqrt{[2 - (-4)]^2 + (2 - 0)^2} = \sqrt{6^2 + 2^2} = \sqrt{36 + 4} = \sqrt{40} = 2\sqrt{10}$

Segments $\overline{AB}$ and $\overline{BC}$ are congruent because they have the same length.

LESSON 1.3 CONTINUED

NAME _____ **DATE** _____

Practice with Examples

For use with pages 17–24

Exercises for Example 2

Find the distance between the points whose coordinates are given.

3. $(6, 4), (-8, 11)$

4. $(-5, 8), (-10, 14)$

5. $(-4, -20), (-10, 15)$

6. $(40, 32), (36, 20)$

7. $(5, -8), (0, 0)$

8. $(a, b), (-a, -b)$

LESSON 1.4

Practice with Examples
For use with pages 26–32

GOAL Use angle postulates and classify angles as acute, right, obtuse, or straight

VOCABULARY

An **angle** consists of two different rays that have the same initial point.

The rays are the **sides** of the angle.

The initial point is the **vertex** of the angle.

Angles that have the same measure are called **congruent angles.**

A point is in the **interior** of an angle if it is between points that lie on each side of the angle.

A point is in the **exterior** of an angle if it is not on the angle or in its interior.

An **acute** angle has measure greater than 0° and less than 90°.

A **right** angle has measure equal to 90°.

An **obtuse** angle has measure greater than 90° and less than 180°.

A **straight** angle has measure equal to 180°.

Two angles are **adjacent angles** if they share a common vertex and side, but have no common interior points.

Postulate 3 Protractor Postulate:
Consider a point A on one side of $\overleftrightarrow{OB}$. The rays of the form $\overrightarrow{OA}$ can be matched one to one with the real numbers from 0 to 180.

The **measure** of $\angle AOB$ is equal to the absolute value of the difference between the real numbers for $\overrightarrow{OA}$ and $\overrightarrow{OB}$.

Postulate 4 Angle Addition Postulate:

If P is in the interior of $\angle RST$, then $m\angle RSP + m\angle PST = m\angle RST$.

EXAMPLE 1 Naming Angles

a. Write three names for the angle and name the vertex and sides of the angle.

b. Suppose R is in the interior of $\angle NOP$, with $m\angle NOR = 23°$ and $m\angle ROP = 27°$. Find $m\angle NOP$.

SOLUTION

a. $\angle NOP$, $\angle PON$, and $\angle O$ are all appropriate names for this angle.

The vertex of this angle is point O and the sides are $\overrightarrow{ON}$ and $\overrightarrow{OP}$.

b. By Angle Addition Postulate, $m\angle NOP = m\angle NOR + m\angle ROP = 23° + 27° = 50°$.

LESSON 1.4 CONTINUED

Practice with Examples
For use with pages 26–32

Exercises for Example 1

Write three names for the angles and name the vertex and sides of each.

1.
2.

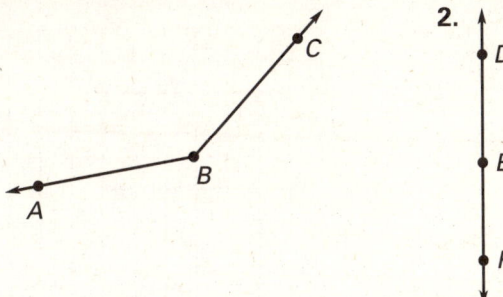

3. Suppose that the angle at the right measures 60° and that there is a point K in the interior of the angle such that $m\angle GHK = 25°$. Find $m\angle KHI$.

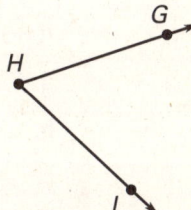

EXAMPLE 2 Classifying Angles in a Coordinate Plane

Plot the points $A(1, 1)$, $B(-1, 1)$, $C(1, 3)$, $D(3, 2)$, and $E(3, 1)$. Then classify the following angles as acute, right, obtuse, or straight.
  a. $\angle CAB$ b. $\angle DAE$ c. $\angle BAD$ d. $\angle EAB$

SOLUTION

Begin by plotting the points, then observe whether each angle is less than 90°, equal to 90°, between 90° and 180°, or equal to 180°.
  a. right angle   b. acute angle
  c. obtuse angle   d. straight angle

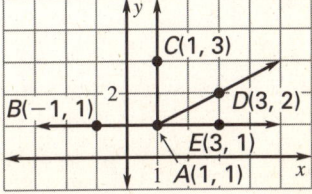

LESSON 1.4 CONTINUED

NAME _____ DATE _____

Practice with Examples
For use with pages 26–32

Exercises for Example 2

Plot the given points and classify the given angles as *acute,* *right,* *obtuse,* **or** *straight*.

4. $A(-2, 4)$, $B(-5, 1)$, $C(0, 0)$, and $D(3, 0)$

   a. $\angle ACB$
   b. $\angle BCD$
   c. $\angle ACD$

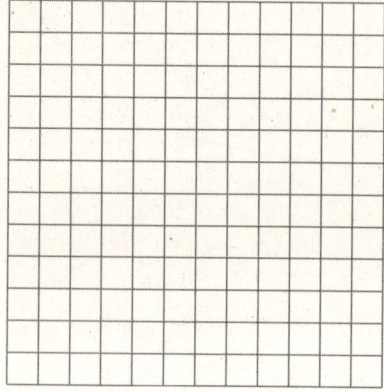

5. $E(4, 0)$, $F(3, 2)$, $G(1, 0)$, $H(-1, -2)$, $I(-1, 2)$

   a. $\angle HGF$
   b. $\angle EGF$
   c. $\angle EGI$
   d. $\angle FGI$
   e. $\angle HGI$

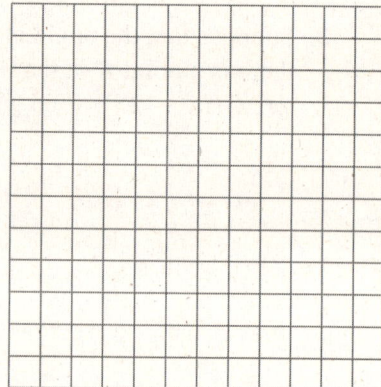

LESSON 1.5

NAME _____ DATE _____

Practice with Examples
For use with pages 34–42

GOAL Bisect a segment and bisect an angle

VOCABULARY

The **midpoint** of a segment is the point that divides, or **bisects,** the segment into two congruent segments.

A **segment bisector** is a segment, ray, line, or plane that intersects a segment at its midpoint.

A **construction** is a geometric drawing that uses a limited set of tools, usually a **compass** and a **straightedge.**

An **angle bisector** is a ray that divides an angle into two adjacent angles that are congruent.

The Midpoint Formula:

If $A(x_1, y_1)$ and $B(x_2, y_2)$ are points in a coordinate plane, then the midpoint of $\overline{AB}$ has coordinates $\left(\dfrac{x_1 + x_2}{2}, \dfrac{y_1 + y_2}{2}\right)$.

EXAMPLE 1 *Finding the Coordinates of the Midpoint of a Segment*

Find the coordinates of the midpoint of $\overline{CD}$ with endpoints $C(4, 3)$ and $D(-2, 0)$.

SOLUTION

Use the Midpoint Formula as follows.

$$M = \left(\dfrac{4 + (-2)}{2}, \dfrac{3 + 0}{2}\right)$$

$$= \left(1, \dfrac{3}{2}\right)$$

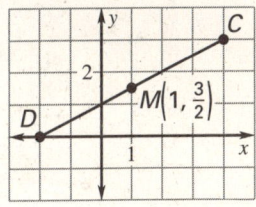

Exercises for Example 1

Find the coordinates of the midpoint of the segment whose endpoints are given.

1. $E(4, -4)$, $F(1, 7)$
2. $G(2, 9)$, $H(-3, 6)$
3. $I(-8, 3)$, $J(3, 0)$

LESSON 1.5 CONTINUED

NAME _____ DATE _____

Practice with Examples
For use with pages 34–42

EXAMPLE 2 Finding the Coordinates of the Endpoint of a Segment

The midpoint of $\overline{KL}$ is $M(6, -2)$. One endpoint is $K(4, 3)$. Find the coordinates of the other endpoint.

SOLUTION

Let (x, y) be the coordinates of L. Use the Midpoint Formula to write equations involving x and y.

$$\frac{4 + x}{2} = 6 \qquad\qquad \frac{3 + y}{2} = -2$$

$$4 + x = 12 \qquad\qquad 3 + y = -4$$

$$x = 8 \qquad\qquad\qquad y = -7$$

So, the other endpoint of the segment is $L(8, -7)$.

Exercises for Example 2

Find the coordinates of the other endpoint of a segment with the given endpoint and midpoint M.

4. $N(-1, 5), M(0, 1)$

5. $P(6, -4), M(3, 10)$

6. $R(-7, -3), M(0, 0)$

LESSON 1.5 CONTINUED

NAME _____ DATE _____

Practice with Examples
For use with pages 34–42

EXAMPLE 3 **Finding the Measure of an Angle**

In the diagram, $\vec{BC}$ bisects $\angle ABD$. Solve for x.

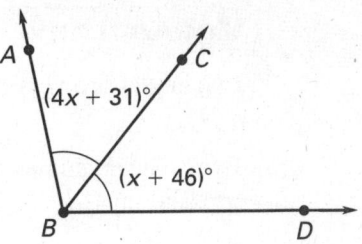

SOLUTION

$m\angle ABC = m\angle CBD$	Congruent angles have equal measures.
$(4x + 31)° = (x + 46)°$	Substitute given measures.
$4x = x + 15$	Subtract 31° from each side.
$3x = 15$	Subtract x from each side.
$x = 5$	Divide each side by 3.

Exercises for Example 3

$\vec{BD}$ bisects $\angle ABC$. Find the value of x.

7.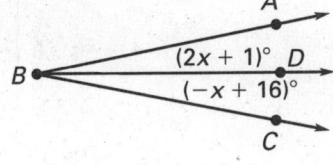

8.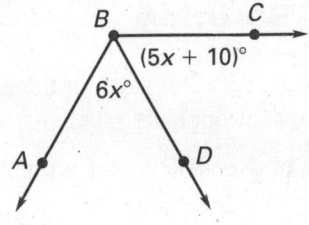

LESSON 1.6

Practice with Examples
For use with pages 44–50

GOAL Identify vertical angles and linear pairs and identify complementary and supplementary angles

> **VOCABULARY**
>
> Two angles are **vertical angles** if their sides form two pairs of opposite rays.
>
> Two adjacent angles are a **linear pair** if their noncommon sides are opposite rays.
>
> Two angles are **complementary angles** if the sum of their measures is 90°. Each angle is the **complement** of the other.
>
> Two angles are **supplementary angles** if the sum of their measures is 180°. Each angle is the **supplement** of the other.

EXAMPLE 1 *Identifying Vertical Angles and Linear Pairs*

a. Are ∠1 and ∠3 vertical angles?

b. Are ∠2 and ∠4 a linear pair?

c. Are ∠1 and ∠4 a linear pair?

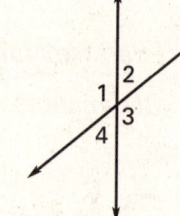

SOLUTION

a. Yes. The sides of the angles form two pairs of opposite rays.

b. No. The angles are not adjacent.

c. Yes. The angles are adjacent and their noncommon sides are opposite rays.

LESSON 1.6 CONTINUED

Practice with Examples
For use with pages 44–50

Exercises for Example 1

Use the figure to answer the questions.

1.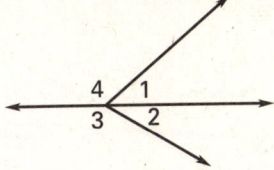

   a. Are ∠1 and ∠2 a linear pair?
   b. Are ∠1 and ∠3 vertical angles?
   c. Are ∠1 and ∠4 a linear pair?
   d. Are ∠2 and ∠4 vertical angles?

2.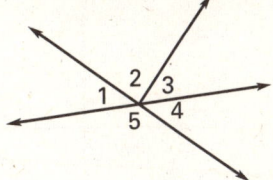

   a. Are ∠1 and ∠5 a linear pair?
   b. Are ∠1 and ∠2 a linear pair?
   c. Are ∠1 and ∠4 vertical angles?
   d. Are ∠3 and ∠5 vertical angles?

EXAMPLE 2 Finding Angle Measures

Solve for x in the diagram at the right. Then find the angle measures.

SOLUTION

Use the fact that vertical angles are congruent.

$$(7x - 25)° = (5x + 15)°$$
$$x = 20$$

Use substitution to find the angle measures.

$m\angle FHI = (7x - 25)° = (7 \cdot 20 - 25)° = 115°$

$m\angle GHJ = (5x + 15)° = (5 \cdot 20 + 15)° = 115°$

Next, realize that ∠FHI and ∠FHG are a linear pair. So, the measures of these two angles must sum to 180°. So, $m\angle FHG = 180° - 115°$, so $m\angle FHG = 65°$.

Finally, notice that ∠FHG and ∠IHJ are vertical angles. So, $m\angle IHJ = 65°$.

LESSON 1.6 CONTINUED

Practice with Examples
For use with pages 44–50

Exercises for Example 2

Solve for *x* and *y*, then find the angle measures.

3.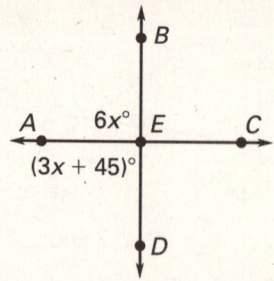

4.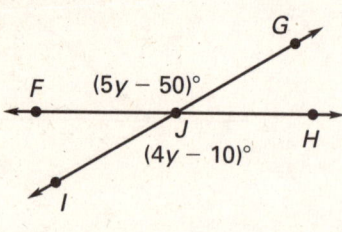

EXAMPLE 3 Finding Measures of Complements and Supplements

a. Given that ∠E is a complement of ∠F and m∠E = 68°, find m∠F.

b. Given that ∠G is a supplement of ∠H and m∠G = 152°, find m∠H.

SOLUTION

a. m∠F = 90° − m∠E = 90° − 68° = 22°

b. m∠H = 180° − m∠G = 180° − 152° = 28°

Exercises for Example 3

Find the measure of the angle.

5. Given that ∠A is a complement of ∠B and m∠B = 81°, find m∠A.

6. Given that ∠C is a supplement of ∠D and m∠C = 27°, find m∠D.

LESSON 1.7

NAME _____ DATE _____

Practice with Examples

For use with pages 51–58

GOAL Find the perimeter and area of common plane figures and use a general problem-solving plan

VOCABULARY

Formulas for the perimeter P, area A, and circumference C of some common plane figures are given below.

Square
Side length s
$P = 4s$
$A = s^2$

Rectangle
length l and width w
$P = 2l + 2w$
$A = lw$

Triangle
Side lengths a, b, and c,
base b, and height h
$P = a + b + c$
$A = \frac{1}{2}bh$

Circle
radius r
$C = 2\pi r$
$A = \pi r^2$

A Problem-Solving Plan:

1. Ask yourself what you need to solve the problem. Write a **verbal model** or **draw a sketch** that will help you find what you need to know.

2. **Label known and unknown facts** on or near your sketch.

3. Use labels and facts to **choose related definitions, theorems, formulas,** or other results you may need.

4. **Reason logically** to link the facts, using a proof or other written argument.

5. Write a **conclusion** that answers the original problem. **Check** that your reasoning is correct.

LESSON 1.7 CONTINUED

Practice with Examples
For use with pages 51–58

EXAMPLE 1 Finding the Perimeter and Area of a Square

Find the perimeter and area of a square with a side of 4 inches.

SOLUTION

Begin by drawing a diagram and labeling one of the sides. Then, use the formulas for perimeter and area of a square.

$P = 4s$ $A = s^2$

$= 4(4)$ $= 4^2$

$= 16$ $= 16$

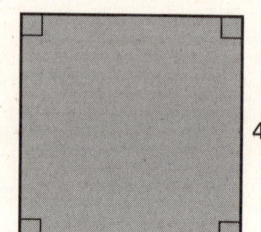

So, the perimeter is 16 inches and the area is 16 square inches.

Exercises for Example 1

Find the perimeter (or circumference) and area of the figure. (Where necessary, use $\pi \approx 3.14$)

1.

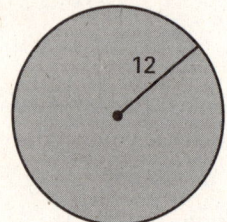

2.

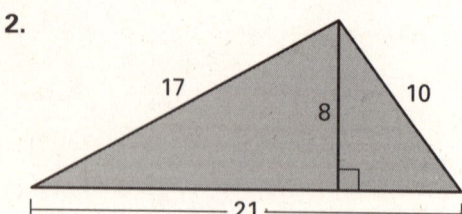

LESSON 1.7 CONTINUED

NAME _____ DATE _____

Practice with Examples
For use with pages 51–58

EXAMPLE 2 Using the Area of a Circle

You are making a cardboard model of a car. You make the tires with a radius of 18 centimeters. If the rim alone has a radius of 14 centimeters, what is the area of the rubber part of the tire?

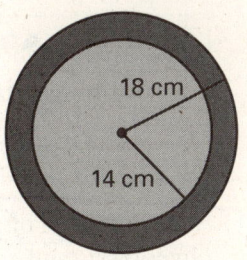

SOLUTION

Draw a Sketch From the diagram, you can see that the area of the rubber can be represented by the area of the larger circle minus the area of the smaller circle.

Verbal Model | Area of rubber | = | Area of large circle | − | Area of small circle |

Labels Area of rubber = A   (square centimeters)

Radius of whole tire = 18   (centimeters)

Radius of rim = 14   (centimeters)

Reasoning $A = \pi \cdot 18^2 - \pi \cdot 14^2$ Write model for rubber area.

$\approx 3.14 \cdot 324 - 3.14 \cdot 196$ $\pi \approx 3.14$ and evaluate powers.

$= 1017.36 - 615.44$ Multiply.

$= 401.92$ Subtract.

The area of the rubber is about 401.92 square centimeters.

Exercises for Example 2

3. A window has the shape of a rectangle with a half-circle (see figure). The rectangle has a width of 3 feet and a height of 7 feet. Find the perimeter and area of the window. Use $\pi \approx 3.14$ where necessary.

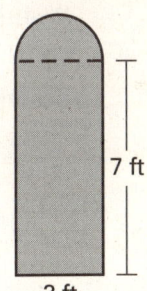

LESSON 2.1

Practice with Examples
For use with pages 71–78

GOAL Recognize and analyze a conditional statement and write postulates about points, lines, and planes using conditional statements

VOCABULARY

A **conditional statement** has two parts, a *hypothesis* and a *conclusion*.

When the statement is written in **if-then form,** the "if" part contains the **hypothesis** and the "then" part contains the **conclusion.**

The **converse** of a conditional statement is formed by switching the hypothesis and conclusion.

A statement can be altered by **negation,** that is, by writing the negative of the statement.

When you negate the hypothesis and conclusion of a conditional statement, you form the **inverse.**

When you negate the hypothesis and conclusion of the converse of a conditional statement, you form the **contrapositive.**

When two statements are both true or both false, they are called **equivalent statements.**

Postulate 5 Through any two points there exists exactly one line.

Postulate 6 A line contains at least two points.

Postulate 7 If two lines intersect, then their intersection is exactly one point.

Postulate 8 Through any three noncollinear points there exists exactly one plane.

Postulate 9 A plane contains at least three noncollinear points.

Postulate 10 If two points lie in a plane, then the line containing them lies in the plane.

Postulate 11 If two planes intersect, then their intersection is a line.

LESSON 2.1 CONTINUED

Practice with Examples
For use with pages 71–78

EXAMPLE 1 — Writing the If-Then Form, Inverse, Converse, and Contrapositive

a. Rewrite the conditional statement in *if-then* form.

b. Write the inverse, converse, and contrapositive of the statement.
   An even number greater than two is not prime.

SOLUTION

a. If a number greater than two is even, then it is not prime.

b. **inverse:** If a number greater than two is not even, then it is prime.
   converse: If a number greater than two is not prime, then it is even.
   contrapositive: If a number greater than two is prime, then it is not even.

Exercises for Example 1

For the conditional statement (a) rewrite it in *if-then* form, (b) write the inverse, (c) write the converse, and (d) write the contrapositive.

1. I will dry the dishes if you will wash them.

2. A square with side of length 3 centimeters has an area of 9 square centimeters.

3. An angle with measure of 90° is a right angle.

LESSON 2.1 CONTINUED

Practice with Examples
For use with pages 71–78

EXAMPLE 2 Using Postulates and Counterexamples

Decide whether the statement is *true* or *false*. If it is false, give a counterexample.

  a. Through any three points there exists exactly one line.

  b. For any one point in a plane, there exists exactly one line through that point in that plane.

SOLUTION

  a. This statement is false. If the three points are noncollinear, then there is no line which includes all three points.

  b. This statement is false. In the diagram at the right, lines *l* and *m* both go through point *A* and both are in plane *P*.

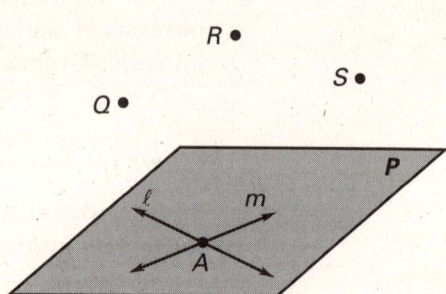

Exercises for Example 2

Decide whether the statement is true or false. If it is false, give a counterexample.

  4. Three planes can intersect in a point.

  5. For noncollinear points *A*, *B*, and *C*, point *C* and the line formed by *A* and *B* are in different planes.

LESSON 2.2

NAME _____ DATE _____

Practice with Examples
For use with pages 79–85

GOAL Recognize and use biconditional statements

> **VOCABULARY**
>
> A **biconditional statement** is a statement that contains the phrase "if and only if."

EXAMPLE 1 Rewriting a Biconditional Statement

Rewrite the biconditional statement as a conditional statement and its converse.

   a. An angle is a right angle if and only if its measure is 90°.

   b. A number is even if and only if it is divisible by two.

   c. A point on a segment is the midpoint of the segment if and only if it bisects the segment.

SOLUTION

   a. **Conditional statement:** If an angle is a right angle, then its measure is 90°.

   Converse: If an angle's measure is 90°, then it is a right angle.

   b. **Conditional statement:** If a number is even, then it is divisible by two.

   Converse: If a number is divisible by two, then it is even.

   c. **Conditional statement:** If a point on a segment is the midpoint of the segment, then it bisects the segment.

   Converse: If a point on a segment bisects the segment, then it is the midpoint of the segment.

Exercises for Example 1

Rewrite the biconditional statement as a conditional statement and its converse.

1. A number is a perfect square if and only if it is the product of some number times itself.

2. Two angles are complementary if and only if the sum of their measures is 90°.

3. A real number is rational if and only if it can be written in the form $\frac{p}{q}$, where p and q are integers and $q \neq 0$.

LESSON 2.2 CONTINUED

Practice with Examples

For use with pages 79–85

EXAMPLE 2 Analyzing a Biconditional Statement

Consider the following statement: You attend school if and only if it is a weekday.

a. Is this a biconditional statement?

b. Is the statement true?

SOLUTION

a. The statement is a biconditional statement because it contains the words "if and only if."

b. The statement can be rewritten as the following statement and its converse.

Conditional statement: If you are in school, then it is a weekday.

Converse: If it is a weekday, then you are in school.

The first of these statements is true, but the second is false (consider weekdays during the summer, or holidays that occur on a weekday). So, the biconditional statement is false.

Exercises for Example 2

Consider the given statement and decide (a) if it is a biconditional statement and (b) if the statement is true.

4. It rains if and only if there are clouds in the sky.

5. Two angles are adjacent angles if and only if they share a vertex and one side but do not have any common interior points.

LESSON 2.2 CONTINUED

Practice with Examples
For use with pages 79–85

EXAMPLE 3 Writing a Biconditional Statement

The following statement is true. Write the converse of the statement and decide whether the converse is *true* or *false*. If the converse is true, combine it with the original statement to form a true biconditional statement. If the converse is false, state a counterexample.

If the product ab is negative, then either a is negative or b is negative.

SOLUTION

Converse: If either a is negative or b is negative, then the product ab is negative.

The converse is false. Consider the counterexample where $a = -3$ but $b = 0$. Then the product ab would be 0, which is not negative.

Exercises for Example 3

Each of the following statements is true. Write the converse of the statement and decide whether the converse is *true* or *false*. If the converse is true, combine it with the original statement to form a true biconditional statement. If the converse is false, state a counterexample.

6. If the sides of two angles form two pairs of opposite rays, then the angles are vertical angles.

7. If the product ab is 0, then either a must be 0 or b must be zero.

LESSON 2.3

Practice with Examples

For use with pages 87–95

GOAL Use symbolic notation to represent logical statements and form conclusions by applying the laws of logic to true statements

VOCABULARY

Law of Detachment

If $p \rightarrow q$ is a true conditional statement and p is true, then q is true.

Law of Syllogism

If $p \rightarrow q$ and $q \rightarrow r$ are true conditional statements, then $p \rightarrow r$ is true.

EXAMPLE 1 Using Symbolic Notation

Let p be "the value of x is 3" and let q be "$x^3 = 27$."

a. Write $p \rightarrow q$ in words. (Note: $\rightarrow$ is read as "implies.")

b. Write $q \rightarrow p$ in words.

c. Decide whether the biconditional statement $p \leftrightarrow q$ is true. (Note: $\leftrightarrow$ is read as "if and only if.")

SOLUTION

a. If the value of x is 3, then $x^3 = 27$.

b. If $x^3 = 27$, then the value of x is 3.

c. The conditional statement in (a) is true and its converse in (b) is also true. So, the biconditional statement $p \leftrightarrow q$ is true.

Exercises for Example 1

Use p and q to (a) write $p \rightarrow q$ in words, (b) write $q \rightarrow p$ in words, and (c) decide whether the biconditional statement $p \leftrightarrow q$ is true.

1. p: two lines in a plane are parallel
   q: two lines in a plane do not intersect

2. p: you are in North America
   q: you are in the United States

LESSON 2.3 CONTINUED

Practice with Examples
For use with pages 87–95

EXAMPLE 2 Writing an Inverse and a Contrapositive

Let p be "a word is misspelled" and q be "a word is not in the dictionary."

  a. Write the contrapositive of $p \to q$.

  b. Write the inverse of $p \to q$.

SOLUTION

  a. Contrapositive: $\sim q \to \sim p$
  If a word is in the dictionary, then it is not misspelled.

  b. Inverse: $\sim p \to \sim q$
  If a word is not misspelled, then it is in the dictionary.

Exercises for Example 2

Use p and q to (a) write the contrapositive of $p \to q$ and (b) write the inverse of $p \to q$.

3. p: the measures of two angles sum to 180°

   q: two angles are supplementary

4. p: a number is divisible by 10

   q: a number is divisible by 5

LESSON 2.3 CONTINUED

Practice with Examples
For use with pages 87–95

EXAMPLE 3   Using the Laws of Deductive Reasoning

Use the true statements to determine whether the conclusion is *true* or *false*.

- If it looks like rain, then I will bring my umbrella to school with me.
- If there are clouds in the sky and the sky is dark, then it looks like rain.
- If I bring my umbrella to school with me, then I will hang it in the classroom closet.
- This morning, there are clouds in the sky and the sky is dark.

Conclusion: My umbrella is hanging in the classroom closet.

SOLUTION

The conclusion is *true*. Let p, q, r, and s represent the following:

p: There are clouds in the sky and the sky is dark.

q: It looks like rain.

r: I bring my umbrella to school with me.

s: My umbrella is hanging in the classroom closet.

p is true, as provided by the fourth point. Because $p \rightarrow q$ is true (second point), $q \rightarrow r$ is true (first point), and $r \rightarrow s$ is true (third point), then $p \rightarrow s$ is true by the Law of Syllogism.

You are given that there are clouds in the sky and that the sky is dark, so p is true. Using the Law of Detachment, you can conclude that my umbrella is hanging in the classroom closet.

Exercises for Example 3

5. Use the true statements to determine whether the conclusion is true or false. Explain your reasoning.
- If Pete's thirsty, he will want to drink a milkshake.
- If Pete wants to drink a milkshake, he will need to use the blender.
- If Pete needs the blender, he will need to clean it first.
- The blender is dirty.

Conclusion: Pete needs to clean the blender.

LESSON 2.4

Practice with Examples
For use with pages 96–101

GOAL Use properties from algebra and use properties of length and measure to justify segment and angle relationships

VOCABULARY

Algebraic Properties of Equality
Let a, b, and c be real numbers.

Addition Property   If $a = b$, then $a + c = b + c$.

Subtraction Property   If $a = b$, then $a - c = b - c$.

Multiplication Property   If $a = b$, then $ac = bc$.

Division Property   If $a = b$ and $c \neq 0$, then $a \div c = b \div c$.

Reflexive Property   For any real number a, $a = a$.

Symmetric Property   If $a = b$, then $b = a$.

Transitive Property   If $a = b$ and $b = c$, then $a = c$.

Substitution Property   If $a = b$, then a can be substituted for b in any equation or expression.

EXAMPLE 1   Writing Reasons

Solve $10 - 2x = 3(x - 2) + 4$ and write a reason for each step.

SOLUTION

$10 - 2x = 3(x - 2) + 4$	Given
$10 - 2x = 3x - 6 + 4$	Distributive property
$10 - 2x = 3x - 2$	Simplify.
$12 - 2x = 3x$	Addition property of equality
$12 = 5x$	Addition property of equality
$\dfrac{12}{5} = x$	Division property of equality

LESSON 2.4 CONTINUED

Practice with Examples
For use with pages 96–101

Exercises for Example 1

Solve the equation and write a reason for each step.

1. $2x + 3 = 7x$

2. $4 + 2(3x + 5) = 11 - x$

3. $6x - 2 = -4(x - 1)$

4. $\frac{1}{5}x + 4 = 2x + \frac{3}{5}$

EXAMPLE 2 Using Properties of Length and Measure

In the diagram, $WY = XZ$.
Show that $WX = YZ$.

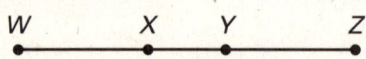

SOLUTION

$WY = XZ$	Given
$WY = WX + XY$	Segment Addition Postulate
$XZ = XY + XZ$	Segment Addition Postulate
$WX + XY = XY + YZ$	Substitution property of equality
$WX = YZ$	Subtraction property of equality

LESSON 2.4 CONTINUED

Practice with Examples

For use with pages 96–101

Exercises for Example 2

Use the given information to show the desired statement.

5. Given that $MN = PQ$, show that $MP = NQ$.

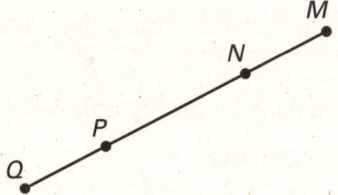

6. Given that $AB = DE$ and $BC = CD$, show that $AD = BE$.

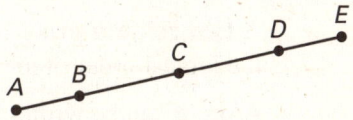

7. Given that $m\angle AQB = m\angle CQD$, show that $m\angle AQC = m\angle BQD$.

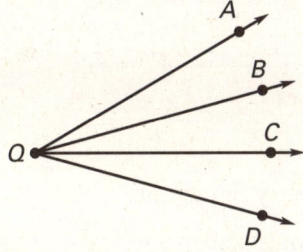

8. Given that $m\angle RPS = m\angle TPV$ and $m\angle TPV = m\angle SPT$, show that $m\angle RPV = 3(m\angle RPS)$

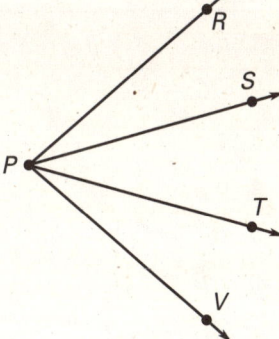

LESSON 2.5

Practice with Examples

For use with pages 102–107

GOAL Write reasons for steps in a proof about segments

VOCABULARY

A true statement that follows as a result of other true statements is called a **theorem.**

A **two-column proof** has numbered statements and reasons that show the logical order of an argument.

A proof can be written in paragraph form, called a **paragraph** proof.

Theorem 2.1 Properties of Segment Congruence

Segment congruence is reflexive, symmetric, and transitive.

EXAMPLE 1 Using Congruence

Use the diagram and the given information to complete the missing steps and reasons in the proof.

Given: $AD = 8$, $BC = 8$, $\overline{BC} \cong \overline{CD}$

Prove: $\overline{AD} \cong \overline{CD}$

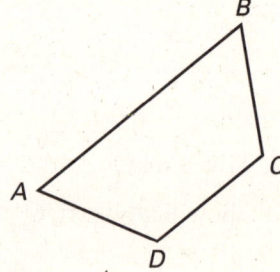

Statements	Reasons
1. a.	1. Given
2. $BC = 8$	2. b.
3. c.	3. Transitive property of equality
4. d.	4. Definition of congruent segments
5. $\overline{BC} \cong \overline{CD}$	5. e.
6. $\overline{AD} \cong \overline{CD}$	6. f.

SOLUTION

a. $AD = 8$ b. Given   c. $AD = BC$ d. $\overline{AD} \cong \overline{BC}$

e. Given   f. Transitive property of congruence

LESSON 2.5 CONTINUED

Practice with Examples
For use with pages 102–107

Exercise for Example 1

Use the diagram and the given information to write a proof.

1. Given: $\overline{BC} \cong \overline{CD}, \overline{AD} \cong \overline{CD}$,
   $AD = 12, AB = 12$
   Prove: $\overline{BC} \cong \overline{BA}$

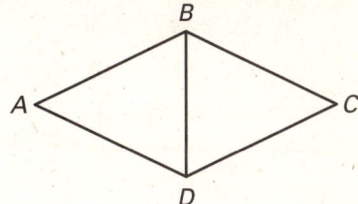

EXAMPLE 2 Using Algebra

Solve for the variable using the given information. Explain your steps.

a. $AC = 91$

b. $\overline{DE} \cong \overline{EF}, \overline{EF} \cong \overline{GF}$

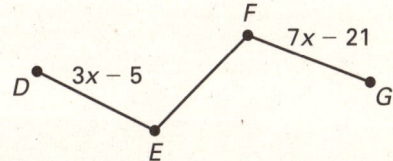

SOLUTION

a. Since the entire segment AC has length 91, the two segments which make up AC must have lengths which add up to 91.

$AB + BC = AC$	Segment Addition Postulate
$x - 10 + 4x + 1 = 91$	Substitution property of equality
$5x - 9 = 91$	Simplify.
$5x = 100$	Addition property of equality
$x = 20$	Division property of equality

b. Because $\overline{DE} \cong \overline{EF}$ and $\overline{EF} \cong \overline{GF}$, $\overline{DE} \cong \overline{GF}$ by the Transitive Property of Congruence. By definition of congruence $DE = GF$.

$3x - 5 = 7x - 21$	Substitution property of equality
$3x + 16 = 7x$	Addition property of equality
$16 = 4x$	Subtraction property of equality
$x = 4$	Division property of equality

LESSON 2.5 CONTINUED

Practice with Examples
For use with pages 102–107

Exercises for Example 2

Solve for the variable using the given information. Explain your steps.

2. Given: $\overline{EG} \cong \overline{HF}$

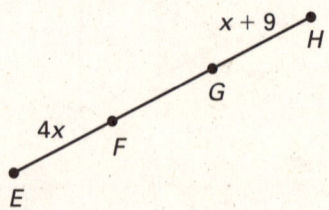

3. Given: $\overline{IP} \cong \overline{JO}, \overline{LM} \cong \overline{IP}$

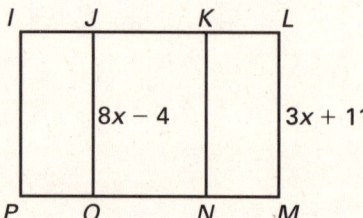

LESSON 2.6

NAME _____ DATE _____

Practice with Examples
For use with pages 109–116

GOAL Use angle congruence properties and prove properties about special pairs of angles

VOCABULARY

Theorem 2.2 Properties of Angle Congruence
Angle congruence is reflexive, symmetric, and transitive.

Theorem 2.3 Right Angle Congruence Theorem
All right angles are congruent.

Theorem 2.4 Congruent Supplements Theorem
If two angles are supplementary to the same angle (or to congruent angles) then they are congruent.

Theorem 2.5 Congruent Complements Theorem
If two angles are complentary to the same angle (or to congruent angles) then the two angles are congruent.

Postulate 12 Linear Pair Postulate
If two angles form a linear pair, then they are supplementary.

Theorem 2.6 Vertical Angles Theorem
Vertical angles are congruent.

EXAMPLE 1 Finding Angles

Complete the statement given that $m\angle AQG = 90°$.

a. $m\angle CQE = ?$

b. If $m\angle BQG = 113°$, then $m\angle EQF = ?$

c. $m\angle AQG + m\angle EQF + m\angle BQC = ?$

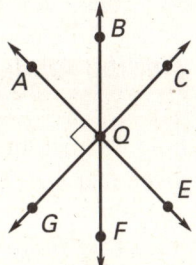

SOLUTION

a. $\angle CQE$ and $\angle AQC$ are vertical angles. By Theorem 2.6, they are congruent. By the definition of congruence, $m\angle CQE = m\angle AQG$, so $m\angle CQE = 90°$.

b. By the Angle Addition Postulate, $m\angle BQG = m\angle AQG + m\angle AQB$. Substituting, you get $113° = 90° + m\angle AQB$, so $m\angle AQB = 23°$ by subtracting. Finally, $\angle EQF \cong \angle AQB$ because they are vertical angles. So, $m\angle EQF = 23°$.

c. $m\angle AQG + m\angle AQB + m\angle BQC = 180°$. $\angle EQF \cong \angle AQB$ because they are vertical angles.
So, $m\angle AQG + m\angle EQF + m\angle BQC = 180°$.

LESSON 2.6 CONTINUED

Practice with Examples
For use with pages 109–116

Exercises for Example 1

Complete the statement given that $m\angle BQD = m\angle CQE = 90°$. Explain your reasoning.

1. $m\angle AQG = ?$

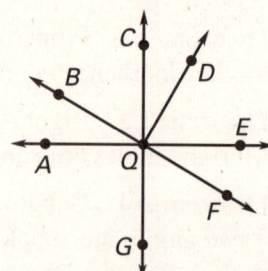

2. $m\angle CQA = ?$

3. If $m\angle CQD = 31°$, then $m\angle EQF = ?$

4. If $m\angle BQG = 125°$, then $m\angle CQF = ?$

5. $m\angle AQB + m\angle GQF + m\angle EQG = ?$

6. If $m\angle EQF = 38°$, then $m\angle BQC = ?$

EXAMPLE 2 Finding Angles

Find the measure of each numbered angle, given that $m\angle DBE = 26°$.

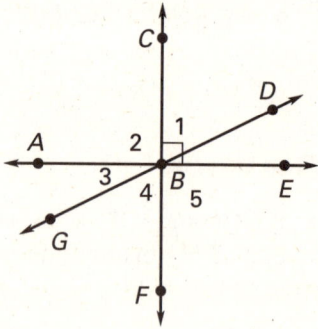

SOLUTION

$\angle CBD$ and $\angle DBE$ are complementary. So, $m\angle 1 = 90° - 26° = 64°$.

$\angle ABC$ and $\angle CBE$ are supplementary. So, $m\angle 2 = 180° - 90° = 90°$.

$\angle ABG$ and $\angle DBE$ are vertical angles. So, $m\angle 3 = 26°$.

$\angle GBF$ and $\angle CBD$ are vertical angles. So, $m\angle 4 = 64°$.

$\angle EBF$ and $\angle CBE$ are supplementary. So, $m\angle 5 = 180° - 90° = 90°$.

LESSON 2.6 CONTINUED

Practice with Examples

For use with pages 109–116

Exercises for Example 2

Find the measure of each indicated angle.

7. ∠ACD and ∠ACB

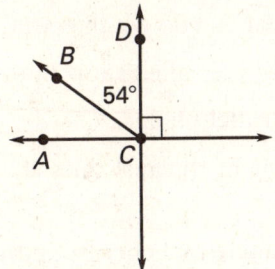

8. ∠QRT and ∠QRU

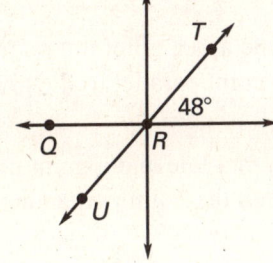

LESSON 3.1

Practice with Examples
For use with pages 129–134

GOAL Identify relationships between lines and identify angles formed by transversals

VOCABULARY

Two lines are **parallel lines** if they are coplanar and do not intersect.

Lines that do not intersect and are not coplanar are called **skew lines.**

Two planes that do not intersect are called **parallel planes.**

A **transversal** is a line that intersects two or more coplanar lines at different points.

When two lines are cut by a transversal, two angles are **corresponding angles** if they occupy corresponding positions.

When two lines are cut by a transversal, two angles are **alternate exterior angles** if they lie outside the two lines on opposite sides of the transversal.

When two lines are cut by a transversal, two angles are **alternate interior angles** if they lie between the two lines on opposite sides of the transversal.

When two lines are cut by a transversal, two angles are **consecutive interior angles** (or **same side interior angles**) if they lie between the two lines on the same side of the transversal.

Postulate 13 *Parallel Postulate* If there is a line and a point not on the line, then there is exactly one line through the point parallel to the given line.

Postulate 14 *Perpendicular Postulate* If there is a line and a point not on the line, then there is exactly one line through the point perpendicular to the given line.

EXAMPLE 1 Identifying Relationships in Space

Think of each segment in the diagram as part of a line. Which of the lines appear to fit the description?

a. parallel to $\overleftrightarrow{AB}$
b. skew to $\overleftrightarrow{AB}$
c. parallel to $\overleftrightarrow{BC}$
d. Are planes *ABE* and *CDE* parallel?

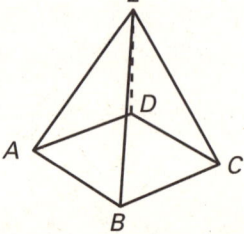

LESSON 3.1 CONTINUED

Practice with Examples
For use with pages 129–134

SOLUTION

a. Only $\overleftrightarrow{CD}$ is parallel to $\overleftrightarrow{AB}$.

b. $\overleftrightarrow{ED}$ and $\overleftrightarrow{EC}$ are skew to $\overleftrightarrow{AB}$.

c. Only $\overleftrightarrow{AD}$ is parallel to $\overleftrightarrow{BC}$.

d. No, the two planes are not parallel. At the very least, we can see that the two planes intersect at point E.

Exercises for Example 1

Think of each segment in the diagram as part of a line. Fill in the blank with *parallel, skew,* or *perpendicular*.

1. $\overleftrightarrow{DE}$ and $\overleftrightarrow{CF}$ are _____.

2. $\overleftrightarrow{AD}$, $\overleftrightarrow{BE}$, and $\overleftrightarrow{CF}$ are _____.

3. Plane ABC and plane DEF are _____.

4. $\overleftrightarrow{BE}$ and $\overleftrightarrow{AB}$ are _____.

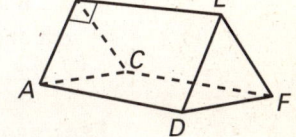

Think of each segment in the diagram as part of a line. There may be more than one right answer.

5. Name a line perpendicular to $\overleftrightarrow{HD}$.

6. Name a plane parallel to DCH.

7. Name a line parallel to $\overleftrightarrow{BC}$.

8. Name a line skew to $\overleftrightarrow{FG}$.

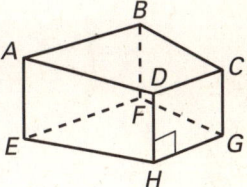

LESSON 3.1 CONTINUED

Practice with Examples
For use with pages 129–134

EXAMPLE 2 Identifying Angle Relationships

List all pairs of angles that fit the description.
a. corresponding
b. alternate exterior
c. alternate interior
d. consecutive interior

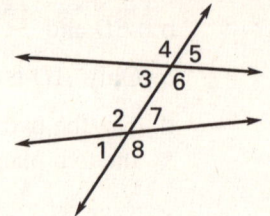

SOLUTION

a. ∠1 and ∠3
   ∠2 and ∠4
   ∠8 and ∠6
   ∠7 and ∠5

b. ∠1 and ∠5
   ∠8 and ∠4

c. ∠2 and ∠6
   ∠7 and ∠3

d. ∠2 and ∠3
   ∠7 and ∠6

Exercises for Example 2

Complete the statement with *corresponding, alternate interior, alternate exterior,* or *consecutive interior.*

9. ∠4 and ∠8 are _____ angles.

10. ∠2 and ∠6 are _____ angles.

11. ∠1 and ∠8 are _____ angles.

12. ∠8 and ∠2 are _____ angles.

13. ∠4 and ∠5 are _____ angles.

14. ∠5 and ∠1 are _____ angles.

LESSON 3.2

Practice with Examples

For use with pages 136–141

GOAL Write different types of proofs and prove results about perpendicular lines

VOCABULARY

A **flow proof** uses arrows to show the flow of the logical argument.

Theorem 3.1 If two lines intersect to form a linear pair of congruent angles, then the lines are perpendicular.

Theorem 3.2 If two sides of two adjacent acute angles are perpendicular, then the angles are complementary.

Theorem 3.3 If two lines are perpendicular, then they intersect to form four right angles.

EXAMPLE 1 Comparing Types of Proofs

Write a two-column proof of Theorem 3.1 (a flow proof is provided in Example 2 on page 137 of the text).

Given:
∠1 ≅ ∠2, ∠1 and ∠2 are a linear pair.

Prove:
$g \perp h$

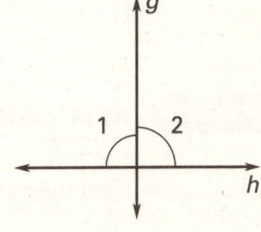

SOLUTION

Statements	Reasons
1. ∠1 ≅ ∠2	1. Given
2. $m\angle 1 = m\angle 2$	2. Definition of congruent angles
3. ∠1 and ∠2 are a linear pair	3. Given
4. ∠1 and ∠2 are supplementary	4. Linear Pair Postulate
5. $m\angle 1 + m\angle 2 = 180°$	5. Definition of supplementary angles
6. $m\angle 1 + m\angle 1 = 180°$	6. Substitution property of equality
7. $2 \cdot (m\angle 1) = 180°$	7. Distributive property
8. $m\angle 1 = 90°$	8. Division property of equality
9. ∠1 is a right ∠	9. Definition of right angle
10. $g \perp h$	10. Definition of perpendicular lines

Geometry
Practice Workbook with Examples

LESSON 3.2 CONTINUED

Practice with Examples
For use with pages 136–141

Exercises for Example 1

1. Write a two-column proof of Theorem 3.2. Note that you are asked to complete a paragraph proof of this theorem in Practice and Applications Exercise 17 on page 139.

2. Write a paragraph proof of Theorem 3.3. Note that you are asked to complete a flow proof related to this theorem in Practice and Applications Exercise 18 on page 139 and a two-column proof related to this theorem in Exercise 19.

EXAMPLE 2 Application of the Theorems

Find the value of x.

a.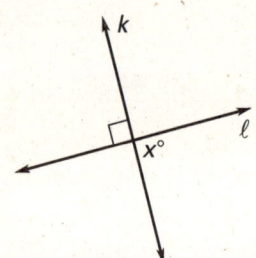

b.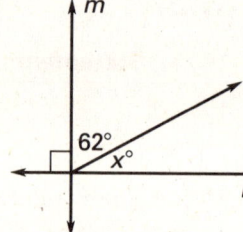

SOLUTION

a. $x = 90$ because, by Theorem 3.3, since k and ℓ are perpendicular, all four angles formed are right angles. By definition of a right angle, x is 90.

b. By Theorem 3.3, since m and n are perpendicular, all four angles formed are right angles. By Theorem 3.2, the 62° angle and the $x°$ angle are complementary. Thus $x + 62 = 90$, so $x = 28$.

LESSON 3.2 CONTINUED

Practice with Examples
For use with pages 136–141

Exercises for Example 2
Find the value of x.

1.
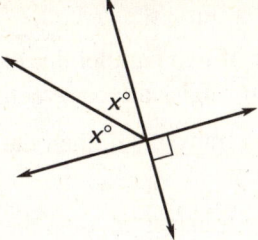

2.
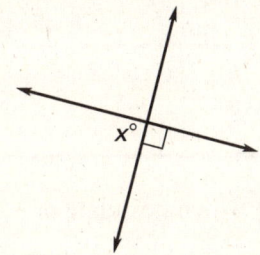

3.
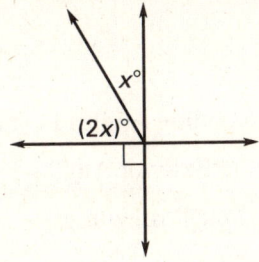

LESSON 3.3

Practice with Examples
For use with pages 143–149

GOAL Prove and use results about parallel lines and transversals and use properties of parallel lines to solve problems

VOCABULARY

Postulate 15 *Corresponding Angles Postulate* If two parallel lines are cut by a transversal, then the pairs of corresponding angles are congruent.

Theorem 3.4 If two parallel lines are cut by a transversal, then the pairs of alternate interior angles are congruent.

Theorem 3.5 If two parallel lines are cut by a transversal, then the pairs of consecutive interior angles are supplementary.

Theorem 3.6 If two parallel lines are cut by a transversal, then the pairs of alternate exterior angles are congruent.

Theorem 3.7 If a transversal is perpendicular to one of two parallel lines, then it is perpendicular to the other.

EXAMPLE 1 Using Properties of Parallel Lines

Given that $m\angle 1 = 32°$, find each measure. Tell which postulate or theorem you use.

a. $m\angle 2$
b. $m\angle 3$
c. $m\angle 4$
d. $m\angle 5$

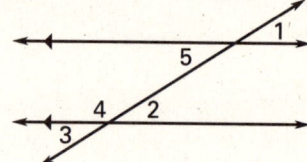

SOLUTION

a. $m\angle 2 = 32°$ Corresponding Angles Postulate
b. $m\angle 3 = 32°$ Alternate Exterior Angles Theorem
c. $m\angle 4 = 180° - m\angle 3 = 148°$ Linear Pair Postulate
d. $m\angle 5 = 32°$ Vertical Angles Theorem

LESSON 3.3 CONTINUED

Practice with Examples
For use with pages 143–149

Exercises for Example 1

Find each measure given that $m\angle 6 = 67°$.

1. $m\angle 7$
2. $m\angle 8$
3. $m\angle 9$
4. $m\angle 10$
5. $m\angle 11$
6. $m\angle 12$
7. $m\angle 13$

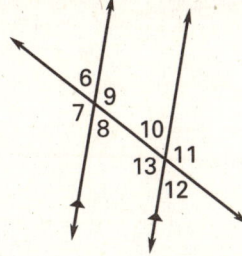

EXAMPLE 2 Using Properties of Parallel Lines

Use properties of parallel lines to find the value of x.

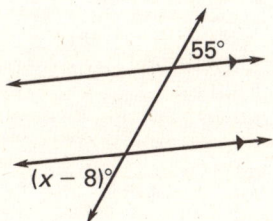

SOLUTION

$(x - 8)° = 55°$ Alternate Exterior Angles Theorem
$x = 63°$ Add.

Exercises for Example 2

Use properties of parallel lines to find the value of x.

8.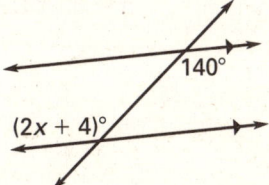

9.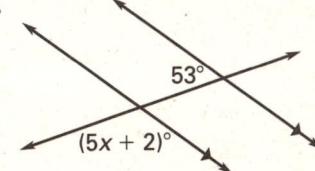

LESSON 3.3 CONTINUED

Practice with Examples
For use with pages 143–149

10.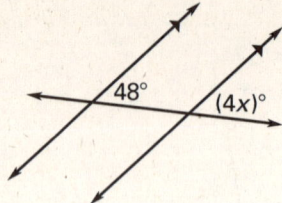

11.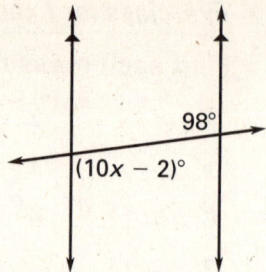

12.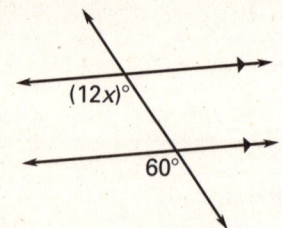

13.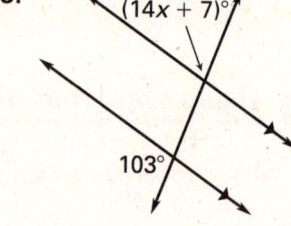

14.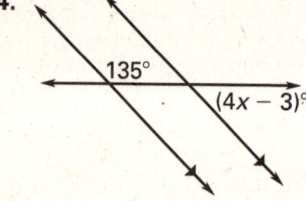

15.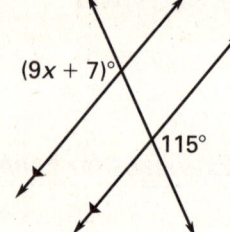

LESSON 3.4

Practice with Examples
For use with pages 150–156

GOAL Prove that two lines are parallel and use properties of parallel lines to solve problems

VOCABULARY

Postulate 16 *Corresponding Angles Converse* If two lines are cut by a transversal so that corresponding angles are congruent, then the lines are parallel.

Theorem 3.8 *Alternate Interior Angles Converse* If two lines are cut by a transversal so that alternate interior angles are congruent, then the lines are parallel.

Theorem 3.9 *Consecutive Interior Angles Converse* If two lines are cut by a transversal so that consecutive interior angles are supplementary, then the lines are parallel.

Theorem 3.10 *Alternate Exterior Angles Converse* If two lines are cut by a transversal so that alternate exterior angles are congruent, then the lines are parallel.

EXAMPLE 1 Proving that Two Lines are Parallel

Prove that lines j and k are parallel.

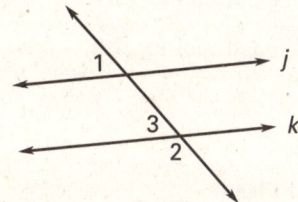

SOLUTION

Given: $m\angle 1 = 53°$
$m\angle 2 = 127°$

Prove: $j \parallel k$

Statements	Reasons
1. $m\angle 1 = 53°$ $m\angle 2 = 127°$	1. Given
2. $m\angle 3 + m\angle 2 = 180°$	2. Linear Pair Postulate
3. $m\angle 3 + 127° = 180°$	3. Substitution prop. of equality
4. $m\angle 3 = 53°$	4. Subtraction prop. of equality
5. $m\angle 3 = m\angle 1$	5. Substitution prop. of equality
6. $\angle 3 \cong \angle 1$	6. Def. of congruent angles
7. $j \parallel k$	7. Corresponding Angles Converse

LESSON 3.4 CONTINUED

Practice with Examples
For use with pages 150–156

Exercises for Example 1

Prove the statement from the given information.

1. Prove: $\ell \parallel m$

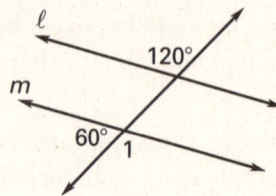

2. Prove: $n \parallel o$

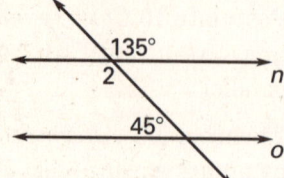

EXAMPLE 2 Identifying Parallel Lines

Determine which rays are parallel.

a. Is $\overrightarrow{PN}$ parallel to $\overrightarrow{SR}$?

b. Is $\overrightarrow{PO}$ parallel to $\overrightarrow{SQ}$?

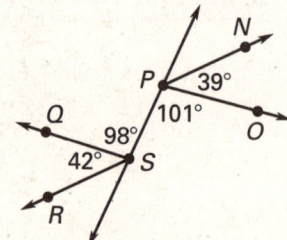

SOLUTION

a. Decide whether $\overrightarrow{PN} \parallel \overrightarrow{SR}$.

$m\angle NPS = 39° + 101°$
$= 140°$
$m\angle RSP = 42° + 98°$
$= 140°$

$\angle NPS$ and $\angle RSP$ are congruent alternate interior angles, so $\overrightarrow{PN} \parallel \overrightarrow{SR}$.

b. Decide whether $\overrightarrow{PO} \parallel \overrightarrow{SQ}$.

$m\angle OPS = 101°$

$m\angle PSQ = 98°$

$\angle OPS$ and $\angle PSQ$ are alternate interior angles, but they are not congruent, so $\overrightarrow{PO}$ and $\overrightarrow{SQ}$ are not parallel.

LESSON 3.4 CONTINUED

Practice with Examples

For use with pages 150–156

Exercises for Example 2

Find the value of x that makes a ∥ b.

3.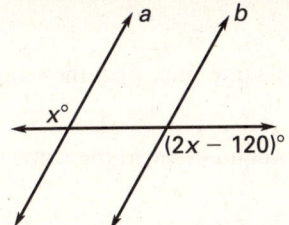

4.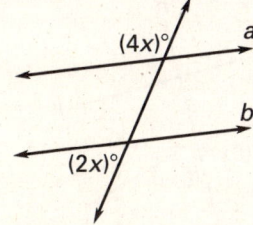

5.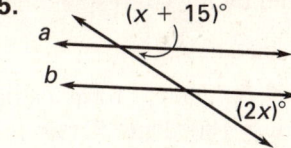

LESSON 3.5

NAME _____ DATE _____

Practice with Examples
For use with pages 157–164

GOAL Use properties of parallel lines and construct parallel lines using straightedge and compass

VOCABULARY

Theorem 3.11 If two lines are parallel to the same line, then they are parallel to each other.

Theorem 3.12 In a plane, if two lines are perpendicular to the same line, then they are parallel to each other.

EXAMPLE 1 Showing Lines are Parallel

Explain how you would show that $k \parallel \ell$.

a.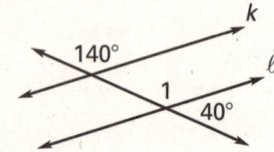
b.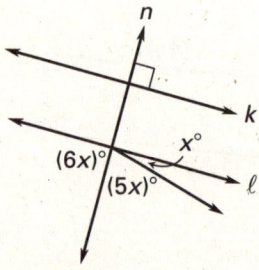

SOLUTION

a. Because the 40° angle and angle 1 form a linear pair, $m\angle 1$ must equal 140°. Thus $\angle 1$ and the other 140° angle are congruent. Because they are also corresponding angles, lines k and ℓ are parallel by the Corresponding Angles Converse postulate.

b. Because the three angles with measures of $6x°$, $5x°$, and $x°$ form a straight line, their sum must be 180°. So $6x + 5x + x = 180$. Thus $12x = 180$, and therefore $x = 15$. We can now conclude that the angle with the $6x°$ measure is a right angle $[6 \cdot (15) = 90]$. Therefore, line n is perpendicular to line ℓ. Since line k is also perpendicular to line n (the 90° angle is indicated), lines k and ℓ are parallel by Theorem 3.12.

52 Geometry
Practice Workbook with Examples

LESSON 3.5 CONTINUED

Practice with Examples
For use with pages 157–164

Exercises for Example 1

Explain how you would show $k \parallel \ell$.

1.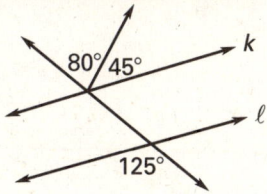

2.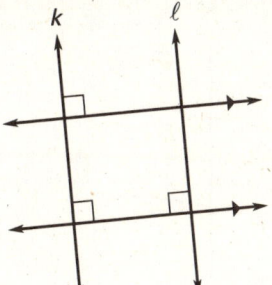

3.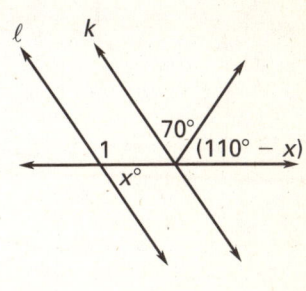

EXAMPLE 2 Naming Parallel Lines

Determine which lines, if any, must be parallel.

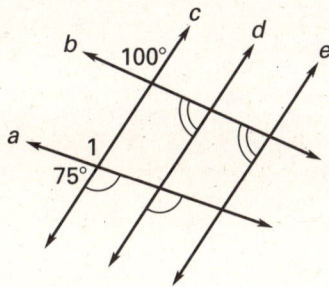

SOLUTION

Lines c and d are parallel because they have congruent corresponding angles. Likewise, lines d and e are parallel because they have congruent corresponding angles. Also, lines c and e are parallel because they are both parallel to the same line, line d. Because $m\angle 1 = 105°$ ($\angle 1$ and the 75° angle form a linear pair), $\angle 1$ and the 100° angle are not congruent. Since $\angle 1$ and the 100° angle are corresponding angles that are not congruent, lines a and b are not parallel.

LESSON 3.5 CONTINUED

NAME _____ DATE _____

Practice with Examples

For use with pages 157–164

Exercises for Example 2

Determine which lines, if any, must be parallel.

4.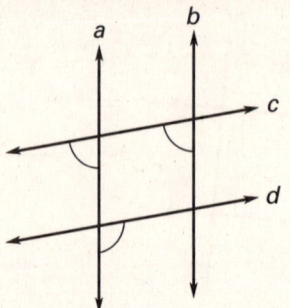

5.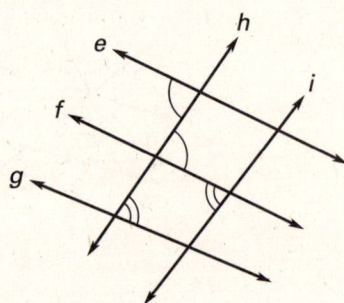

6.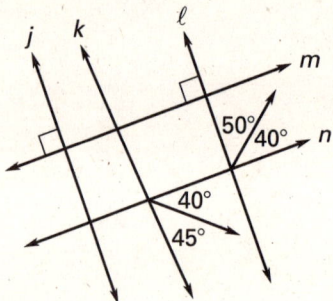

LESSON 3.6

NAME _____ DATE _____

Practice with Examples

For use with pages 165–171

GOAL Find slopes of lines and use slope to identify parallel lines in a coordinate plane and write equations of parallel lines in a coordinate plane

VOCABULARY

Postulate 17 *Slopes of Parallel Lines* In a coordinate plane, two nonvertical lines are parallel if and only if they have the same slope. Any two vertical lines are parallel.

EXAMPLE 1 Finding the Slope of a Line

Find the slope of the line that passes through the points $(3, -3)$ and $(0, 9)$.

SOLUTION

Let $(x_1, y_1) = (3, -3)$ and $(x_2, y_2) = (0, 9)$.

$m = \dfrac{y_2 - y_1}{x_2 - x_1}$

$= \dfrac{9 - (-3)}{0 - 3}$

$= \dfrac{12}{-3}$

$= -4$

The slope of the line is -4.

Exercises for Example 1

Find the slope of the line that passes through the given points.

1. $(4, 2)$ and $(6, 8)$
2. $(-3, -1)$ and $(-5, -11)$
3. $(-8, 12)$ and $(0, -12)$

4. $(8, 3)$ and $(14, 5)$
5. $(-7, -5)$ and $(5, 4)$
6. $(-18, 5)$ and $(4, 5)$

LESSON 3.6 CONTINUED

NAME _____ DATE _____

Practice with Examples

For use with pages 165–171

EXAMPLE 2 Identifying Parallel Lines

Find the slope of each line. Is $a \parallel b$?

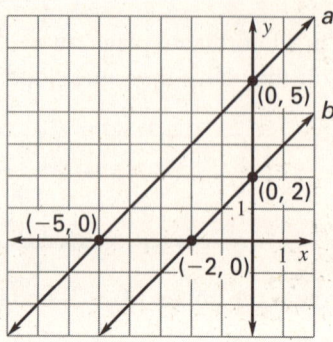

SOLUTION

Find the slope of a. Line a passes through $(-5, 0)$ and $(0, 5)$.

$$m_a = \frac{5 - 0}{0 - (-5)} = \frac{5}{5} = 1$$

Find the slope of b. Line b passes through $(-2, 0)$ and $(0, 2)$.

$$m_b = \frac{2 - 0}{0 - (-2)} = \frac{2}{2} = 1$$

Compare the slopes. Because a and b have the same slope, they are parallel.

Exercises for Example 2

Find the slope of each line. Which lines are parallel?

7.
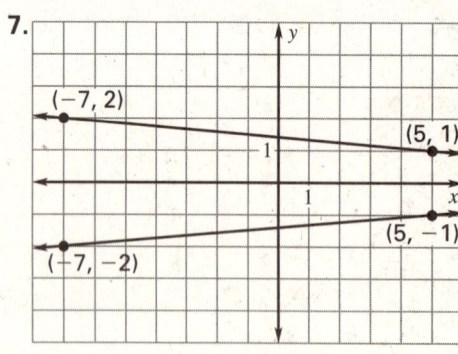

8.
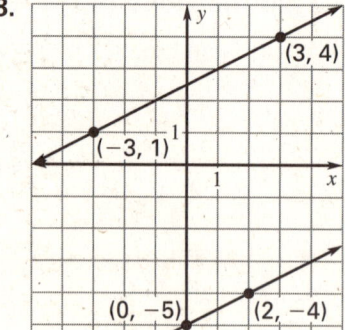

LESSON 3.6 CONTINUED

NAME _____ DATE _____

Practice with Examples

For use with pages 165–171

EXAMPLE 3 Writing an Equation of a Parallel Line

Line k has the equation $y = -x - 4$.

Line ℓ is parallel to k and passes through the point $(1, 5)$. Write an equation of ℓ.

SOLUTION

Find the slope. The slope of k is -1. Because parallel lines have the same slope, the slope of ℓ is also -1.

Solve for b. Use $(x, y) = (1, 5)$ and $m = -1$.

$y = mx + b$

$5 = -1(1) + b$

$5 = -1 + b$

$6 = b$

Write an equation. Because $m = -1$ and $b = 6$, an equation of ℓ is $y = -x + 6$.

Exercises for Example 3

Write an equation of the line the passes through the given point P and is parallel to the line with the given equation.

9. $P(10, 3), y = x - 12$

10. $P(-5, 2), y = -x - 9$

11. $P(-1, 2), y = \frac{2}{3}x - 2$

LESSON 3.7

Practice with Examples
For use with pages 172–178

GOAL Use slope to identify perpendicular lines in a coordinate plane and write equations of perpendicular lines

VOCABULARY

Postulate 18 *Slopes of Perpendicular Lines* In a coordinate plane, two nonvertical lines are perpendicular if and only if the product of their slopes is -1.

EXAMPLE 1 Deciding Whether Lines are Perpendicular

a. Decide whether $\overleftrightarrow{PQ}$ and $\overleftrightarrow{QR}$ are perpendicular.

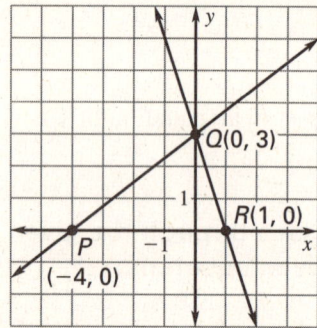

b. Decide whether the lines are perpendicular.

Line ℓ: $2x - 3y = -4$ Line k: $3x + 2y = 3$

SOLUTION

a. Find each slope.

Slope of $\overleftrightarrow{PQ} = \dfrac{3 - 0}{0 - (-4)} = \dfrac{3}{4}$

Slope of $\overleftrightarrow{QR} = \dfrac{0 - 3}{1 - 0} = \dfrac{-3}{1} = -3$

Multiply slopes to see if the lines are perpendicular.

$\dfrac{3}{4} \cdot (-3) = -\dfrac{9}{4}$

The product of the slopes is not -1. So, $\overleftrightarrow{PQ}$ and $\overleftrightarrow{QR}$ are not perpendicular.

LESSON 3.7 CONTINUED

Practice with Examples
For use with pages 172–178

b. Rewrite each equation in slope-intercept form to find the slope.

Line ℓ:
$y = \dfrac{2}{3}x + \dfrac{4}{3}$
slope $= \dfrac{2}{3}$

Line k:
$y = -\dfrac{3}{2}x + \dfrac{3}{2}$
slope $= -\dfrac{3}{2}$

Multiply the slopes to see if the lines are perpendicular.

$\left(\dfrac{2}{3}\right) \cdot \left(-\dfrac{3}{2}\right) = -1$, so the lines are perpendicular.

Exercises for Example 1

Decide whether lines k and ℓ are perpendicular.

1. k passes through $(3, 2)$ and $(-1, 5)$

    ℓ passes through $(0, 2)$ and $(3, 6)$

2. k has the equation $2x - 4y = -3$

    ℓ has the equation $x + 2y = -6$

LESSON 3.7 CONTINUED

NAME _____ DATE _____

Practice with Examples

For use with pages 172–178

EXAMPLE 2 Writing the Equation of a Perpendicular Line

Line k has equation $y = \frac{2}{3}x - \frac{4}{3}$. Find an equation of line ℓ that passes through $P(3, -1)$ and is perpendicular to k.

SOLUTION

First determine the slope of ℓ. For k and ℓ to be perpendicular, the product of their slopes must equal -1.

$m_k \cdot m_\ell = -1$

$\frac{2}{3} \cdot m_\ell = -1$

$m_\ell = -\frac{3}{2}$

Then use $m = -\frac{3}{2}$ and $(x, y) = (3, -1)$ to find b.

$y = mx + b$

$-1 = -\frac{3}{2} \cdot (3) + b$

$\frac{7}{2} = b$

So, an equation of ℓ is $y = -\frac{3}{2}x + \frac{7}{2}$.

Exercises for Example 2

Line j is perpendicular to the line with the given equation and line j passes through P. Write an equation of line j.

3. $4x + 7y = 13$, $P(-2, 6)$

4. $5x - 2y = 3$, $P\left(0, -\frac{3}{2}\right)$

5. $x + 5y = 6$, $P(-1, 2)$

LESSON 4.1

Practice with Examples
For use with pages 194–201

GOAL Classify triangles by their sides and angles and find angle measures in triangles

> **VOCABULARY**
>
> A **triangle** is a figure formed by three segments joining three non-collinear points.
>
> An **equilateral triangle** has three congruent sides.
>
> An **isosceles triangle** has at least two congruent sides.
>
> A **scalene triangle** has no congruent sides.
>
> An **acute triangle** has three acute angles.
>
> An **equiangular triangle** has three congruent angles.
>
> A **right triangle** has one right angle.
>
> An **obtuse triangle** has one obtuse angle.
>
> The three angles of a triangle are the **interior angles**.
>
> When the sides of a triangle are extended, the angles that are adjacent to the interior angles are **exterior angles.**
>
> **Theorem 4.1 Triangle Sum Theorem**
> The sum of the measures of the interior angles of a triangle is 180°.
>
> **Theorem 4.2 Exterior Angle Theorem**
> The measure of an exterior angle of a triangle is equal to the sum of the measures of the two nonadjacent interior angles.
>
> **Corollary to the Triangle Sum Theorem**
> The acute angles of a right triangle are complementary.

LESSON 4.1 CONTINUED

Practice with Examples

For use with pages 194–201

EXAMPLE 1 — Classifying Triangles

Classify the triangles by their sides and angles.

a.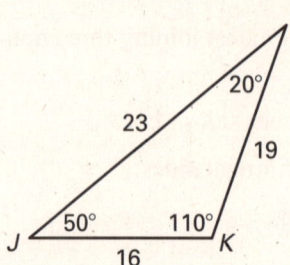

b.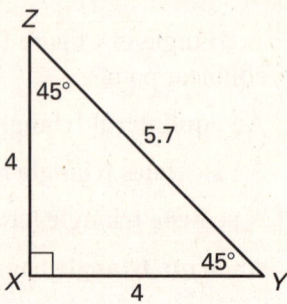

SOLUTION

a. $\triangle JKL$ has one obtuse angle and no congruent sides. It is an obtuse scalene triangle.

b. $\triangle XYZ$ has one right angle and two congruent sides. It is a right isosceles triangle.

Exercises for Example 1

Classify the triangle by its sides and angles.

1.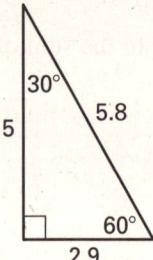

2.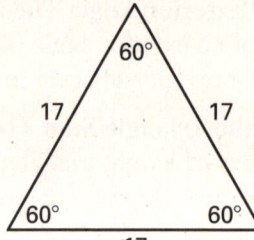

3.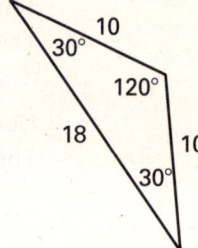

NAME ______________________ DATE __________

Practice with Examples

For use with pages 194–201

EXAMPLE 2 Finding Angle Measures

a. Find the value of x.

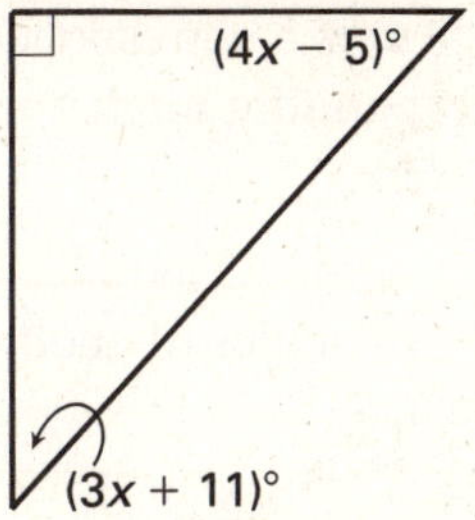

b. Find the value of y.

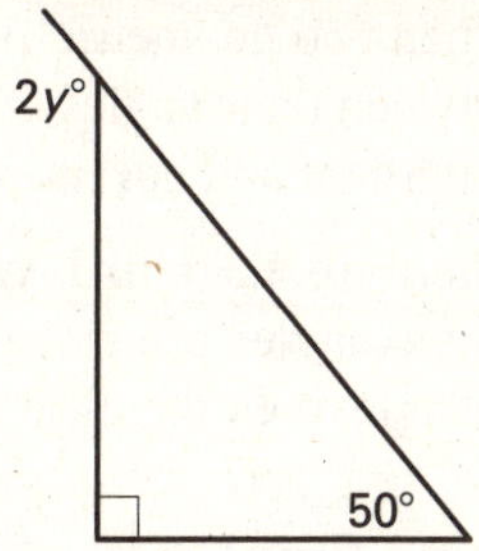

Solution

a. From the Corollary to the Triangle Sum Theorem, you can write and solve an equation to find the value of x.

$(4x - 5)^\circ + (3x + 11)^\circ = 90^\circ$ — The acute angles of a right triangle are complementary.

$x = 12$ — Solve for x.

b. You can apply the Exterior Angle Theorem to write and solve an equation that will allow you to find the value of y.

$90^\circ + 50^\circ = 2y^\circ$ — Apply the Exterior Angle Theorem.

$y = 70$ — Solve for y.

Exercises for Example 2

Find the value of x.

4.

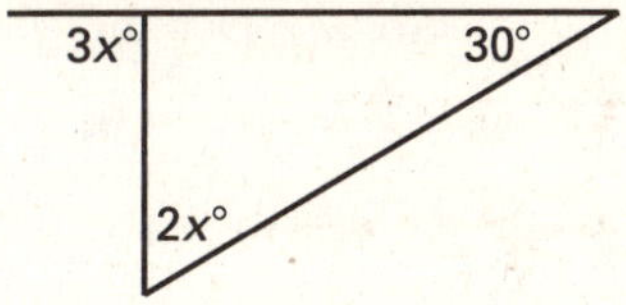

5.

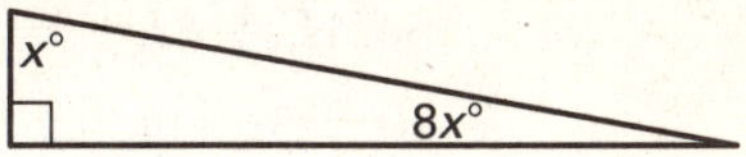

LESSON 4.2

Practice with Examples
For use with pages 202–210

GOAL Identify congruent figures and corresponding parts

VOCABULARY

When two geometric figures are **congruent**, there is a correspondence between their angles and sides such that **corresponding angles** are congruent and **corresponding sides** are congruent.

Theorem 4.3 Third Angles Theorem
If two angles of one triangle are congruent to two angles of another triangle, then the third angles are also congruent.

EXAMPLE 1 Using Properties of Congruent Figures

In the diagram, $ABCDE \cong FGHIJ$

a. Find the value of x.
b. Find the value of y.

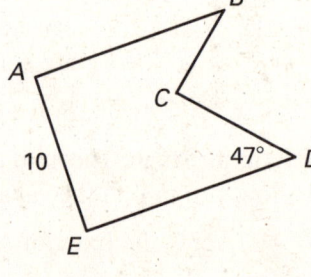
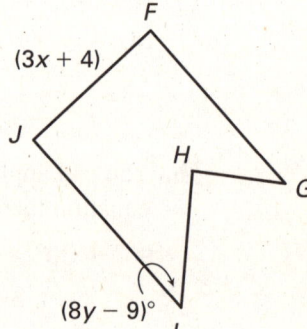

SOLUTION

a. You know that $\overline{AE} \cong \overline{FJ}$.
   So $AE = FJ$.
   $10 = 3x + 4$
   $x = 2$

b. You know that $\angle D \cong \angle I$.
   So, $m\angle D = m\angle I$.
   $47° = (8y - 9)°$
   $56 = 8y$
   $y = 7$

LESSON 4.2 CONTINUED

NAME _____ DATE _____

Practice with Examples
For use with pages 202–210

Exercises for Example 1

In Exercises 1 and 2, for each pair of figures find (a) the value of *x* and (b) the value of *y*.

1. △ABC ≅ △DEF

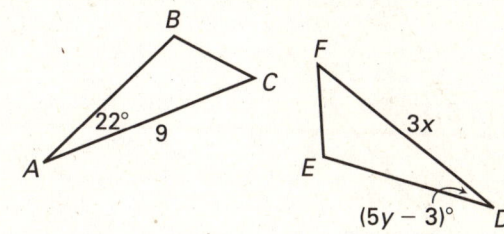

2. ABCDEF ≅ GHIJKL

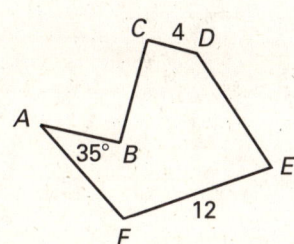
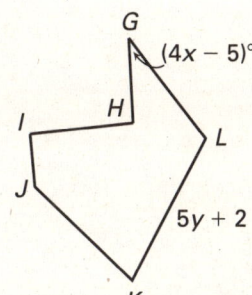

EXAMPLE 2 Using the Third Angles Theorem

Find the value of *x*.

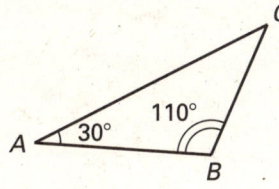
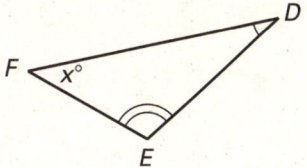

SOLUTION

In the diagram, ∠A ≅ ∠D and ∠B ≅ ∠E. From the Third Angles Theorem, you know that ∠C ≅ ∠F. So, m∠C = m∠F.

From the Triangle Sum Theorem, m∠C = 180° − 30° − 110° = 40°.

m∠C = m∠F   Third Angles Theorem
40 = x   Substitute.

Practice with Examples

For use with pages 202–210

Exercises for Example 2

Find the value of x.

3.

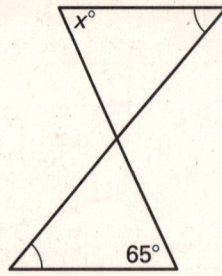

4.

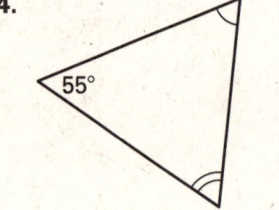 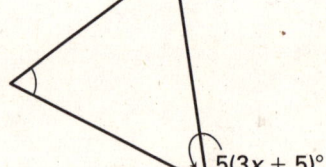

LESSON 4.3

Practice with Examples
For use with pages 212–219

GOAL Prove that triangles are congruent using the SSS and SAS Congruence Postulates

> **Postulate 19 Side-Side-Side (SSS) Congruence Postulate**
> If three sides of one triangle are congruent to three sides of a second triangle, then the two triangles are congruent.
>
> **Postulate 20 Side-Angle-Side (SAS) Congruence Postulate**
> If two sides and the included angle of one triangle are congruent to two sides and the included angle of a second triangle, then the two triangles are congruent.

EXAMPLE 1 — Using the SAS Congruence Postulate

Prove that △ABC ≅ △DEF.

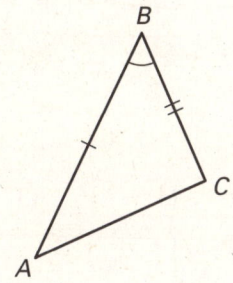
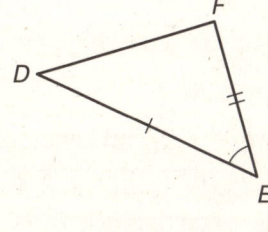

SOLUTION

The marks on the diagram show that $\overline{AB} \cong \overline{DE}$, $\overline{BC} \cong \overline{EF}$, and ∠B ≅ ∠E. So, by the SAS Congruence Postulate, you know that △ABC ≅ △DEF.

Exercises for Example 1

State the congruence postulate you would use to prove that the two triangles are congruent.

1.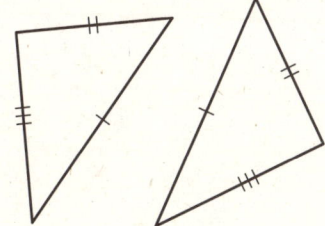

2.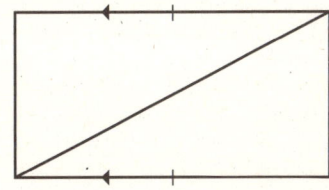

LESSON 4.3 CONTINUED

NAME _____ DATE _____

Practice with Examples
For use with pages 212–219

EXAMPLE 2 **Congruent Triangles in a Coordinate Plane**

Use the SSS Congruence Postulate to show that $\triangle ABC \cong \triangle CDE$.

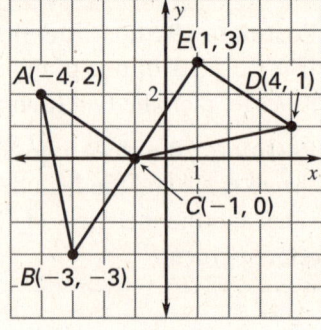

SOLUTION

Use the distance formula to show that corresponding sides are the same length. For all lengths, $d = \sqrt{(x_2 - x_1)^2 + (y_2 - y_1)^2}$.

$AB = \sqrt{(-3 - (-4))^2 + (-3 - 2)^2}$
$= \sqrt{1^2 + (-5)^2}$
$= \sqrt{26}$

$CD = \sqrt{(4 - (-1))^2 + (1 - 0)^2}$
$= \sqrt{5^2 + 1^2}$
$= \sqrt{26}$

So, $AB = CD$, and hence $\overline{AB} \cong \overline{CD}$.

$BC = \sqrt{(-1 - (-3))^2 + (0 - (-3))^2}$
$= \sqrt{2^2 + 3^2}$
$= \sqrt{13}$

$DE = \sqrt{(1 - 4)^2 + (3 - 1)^2}$
$= \sqrt{(-3)^2 + 2^2}$
$= \sqrt{13}$

So, $BC = DE$, and hence $\overline{BC} \cong \overline{DE}$.

$CA = \sqrt{(-4 - (-1))^2 + (2 - 0)^2}$
$= \sqrt{(-3)^2 + 2^2}$
$= \sqrt{13}$

$EC = \sqrt{(-1 - 1)^2 + (0 - 3)^2}$
$= \sqrt{(-2)^2 + (-3)^2}$
$= \sqrt{13}$

So, $CA = EC$, and hence $\overline{CA} \cong \overline{EC}$.

So, by the SSS Congruence Postulate, you know that $\triangle ABC \cong \triangle CDE$.

LESSON 4.3 CONTINUED

NAME _____ DATE _____

Practice with Examples

For use with pages 212–219

Exercise for Example 2

3. Prove that $\triangle ABC \cong \triangle DEF$.

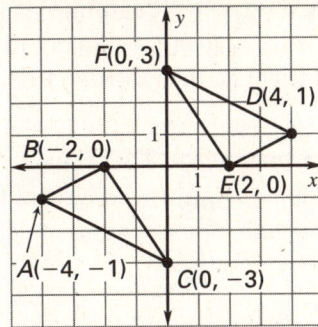

LESSON 4.4

NAME _____ DATE _____

Practice with Examples
For use with pages 220–227

GOAL Prove that triangles are congruent using the ASA Congruence Postulate and the AAS Congruence Theorem

> **Postulate 21 Angle-Side-Angle (ASA) Congruence Postulate**
> If two angles and the included side of one triangle are congruent to two angles and the included side of a second triangle, then the two triangles are congruent.
>
> **Theorem 4.5 Angle-Angle-Side (AAS) Congruence Theorem**
> If two angles and a nonincluded side of one triangle are congruent to two angles and the corresponding nonincluded side of a second triangle, then the two triangles are congruent.

EXAMPLE 1 Proving Triangles are Congruent Using the ASA Congruence Postulate

Given: $\overline{BC} \cong \overline{EC}$, $\angle B \cong \angle E$
Prove: $\triangle ABC \cong \triangle DEC$

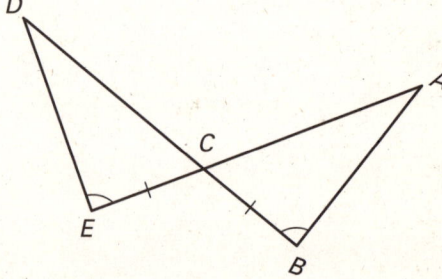

SOLUTION

Statements	Reasons
1. $\overline{BC} \cong \overline{EC}$	1. Given
2. $\angle B \cong \angle E$	2. Given
3. $\angle ACB \cong \angle DCE$	3. Vertical Angles Theorem
4. $\triangle ABC \cong \triangle DEC$	4. ASA Congruence Postulate

Geometry
Practice Workbook with Examples

LESSON 4.4 CONTINUED

Practice with Examples
For use with pages 220–227

Exercises for Example 1

In Exercises 1 and 2, use the given information to prove that the triangles are congruent.

1. Given: $\overline{MC} \cong \overline{AC}$
   $\angle NMC$ and $\angle BAC$ are right angles.
   Prove: $\triangle NMC \cong \triangle BAC$

2. Given: $\overline{AE} \cong \overline{DE}$, $\angle A \cong \angle D$
   Prove: $\triangle BAE \cong \triangle CDE$

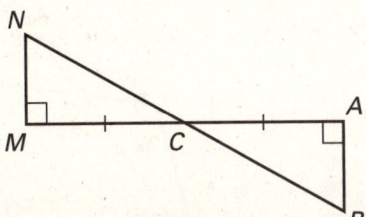
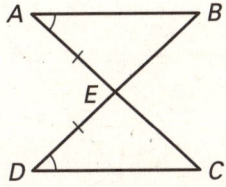

EXAMPLE 2 Proving Triangles are Congruent Using AAS Congruence Theorem

Given: $\overline{AD} \cong \overline{AE}$, $\angle B \cong \angle C$
Prove: $\triangle ABD \cong \triangle ACE$

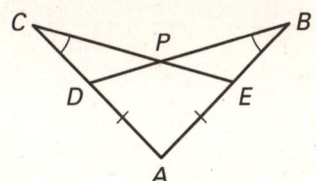

SOLUTION

Statements	Reasons
1. $\overline{AD} \cong \overline{AE}$	1. Given
2. $\angle B \cong \angle C$	2. Given
3. $\angle A \cong \angle A$	3. Reflexive Property of Congruence
4. $\triangle ABD \cong \triangle ACE$	4. AAS Congruence Theorem

LESSON 4.4 CONTINUED

Practice with Examples

For use with pages 220–227

Exercises for Example 2

In Exercises 3 and 4, use the given information to prove that the triangles are congruent.

3. Given: $\angle G \cong \angle B$, $\overline{CB} \parallel \overline{GA}$

   Prove: $\triangle GCA \cong \triangle BAC$

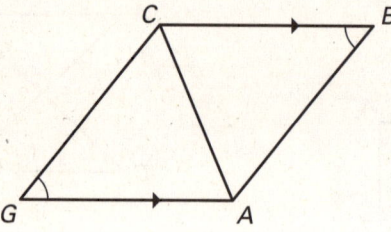

4. Given: $\angle OMN \cong \angle ONM$, $\angle LMO \cong \angle JNO$

   Prove: $\triangle MJN \cong \triangle NLM$

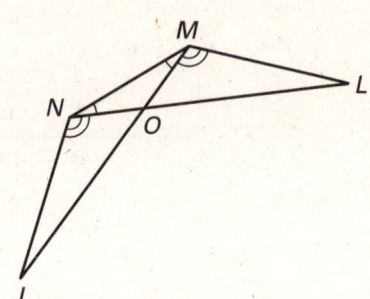

LESSON 4.5

Practice with Examples
For use with pages 229–235

GOAL Use congruent triangles to plan and write proofs

EXAMPLE 1 Planning and Writing a Proof

Given: $\overline{PR} \cong \overline{PQ}$, $\overline{SR} \cong \overline{TQ}$

Prove: $\overline{QS} \cong \overline{RT}$

Plan for Proof: $\overline{QS}$ and $\overline{RT}$ are corresponding parts of $\triangle PQS$ and $\triangle PRT$ and also of $\triangle RQS$ and $\triangle QRT$.

The first set of triangles is easier to prove congruent than the second set. Then use the fact that corresponding parts of congruent triangles are congruent.

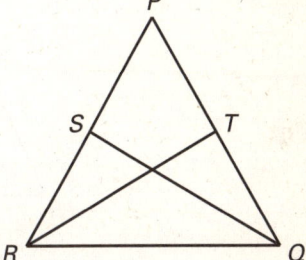

SOLUTION

Statements	Reasons
1. $\overline{PR} \cong \overline{PQ}$	1. Given
2. $PR = PQ$	2. Definition of congruence
3. $PR = PS + SR$	3. Segment Addition Postulate
4. $PQ = PT + TQ$	4. Segment Addition Postulate
5. $PS + SR = PT + TQ$	5. Subsitution
6. $\overline{SR} \cong \overline{TQ}$	6. Given
7. $SR = TQ$	7. Definition of congruence
8. $PS = PT$	8. Subtraction property of equality
9. $\overline{PS} \cong \overline{PT}$	9. Definition of congruence
10. $\angle P \cong \angle P$	10. Reflexive Property of Congruence
11. $\triangle PQS \cong \triangle PRT$	11. SAS Congruence Postulate
12. $\overline{QS} \cong \overline{RT}$	12. Corresponding parts of congruent triangles are congruent.

LESSON 4.5 CONTINUED

Practice with Examples

For use with pages 229–235

Exercises for Example 1

Use the given information to prove the desired statement.

1. Given: $\overline{RT} \cong \overline{AS}$, $\overline{RS} \cong \overline{AT}$
   Prove: $\angle TSA \cong \angle STR$

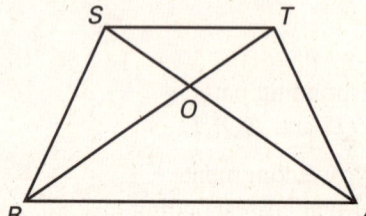

2. Given: $\angle 1 \cong \angle 2 \cong \angle 3$, $\angle 4 \cong \angle 5$, $\overline{ES} \cong \overline{DT}$
   Prove: $\overline{HE} \cong \overline{HD}$

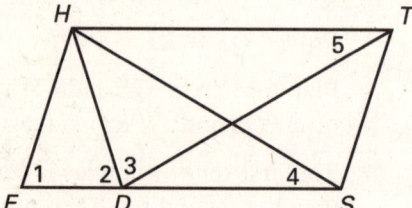

EXAMPLE 2 — Using More than One Pair of Triangles

Given: $\angle 1 \cong \angle 2$, $\angle 5 \cong \angle 6$

Prove: $\overline{AC} \perp \overline{BD}$

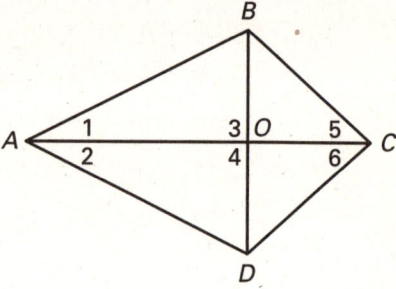

Plan for Proof: It can be helpful to reason backward from what is to be proved. You can show that $\overline{AC} \perp \overline{BD}$ if you can show $\angle 3 \cong \angle 4$. Notice that $\angle 3$ and $\angle 4$ are corresponding parts of $\triangle ABO$ and $\triangle ADO$. You can prove $\triangle ABO \cong \triangle ADO$ by SAS if you first prove $\overline{AB} \cong \overline{AD}$. $\overline{AB}$ and $\overline{AD}$ are corresponding parts of $\triangle ABC$ and $\triangle ADC$. You can prove $\triangle ABC \cong \triangle ADC$ by ASA.

LESSON 4.5 CONTINUED

Practice with Examples
For use with pages 229–235

SOLUTION

Statements	Reasons
1. $\angle 1 \cong \angle 2$, $\angle 5 \cong \angle 6$	1. Given
2. $\overline{AC} \cong \overline{AC}$	2. Reflexive Property of Congruence
3. $\triangle ABC \cong \triangle ADC$	3. ASA Congruence Postulate
4. $\overline{AB} \cong \overline{AD}$	4. Corresponding parts of congruent triangles are congruent.
5. $\overline{AO} \cong \overline{AO}$	5. Reflexive Property of Congruence
6. $\triangle ABO \cong \triangle ADO$	6. SAS Congruence Postulate
7. $\angle 3 \cong \angle 4$	7. Corresponding parts of congruent triangles are congruent.
8. $\overline{AC} \perp \overline{BD}$	8. If 2 lines intersect to form a linear pair of congruent angles, then the lines are $\perp$.

Exercises for Example 2

In Exercises 3 and 4, use the given information to prove the desired statement.

3. Given: $\overline{PA} \cong \overline{KA}$, $\overline{LA} \cong \overline{NA}$
   Prove: $\overline{AX} \cong \overline{AY}$

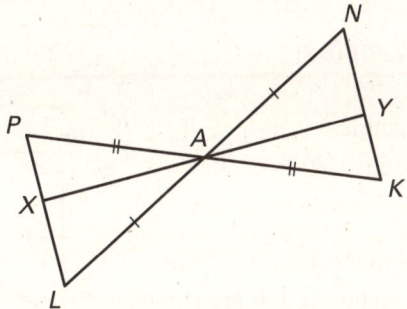

4. Given: $\angle DAL \cong \angle BCM$,
   $\angle CDL \cong \angle ABM$
   $\overline{DC} \cong \overline{BA}$
   Prove: $\overline{AL} \cong \overline{CM}$

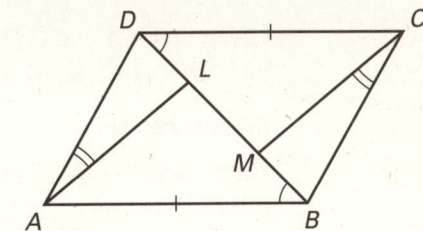

LESSON 4.6

NAME _____ DATE _____

Practice with Examples
For use with pages 236–242

GOAL Use properties of isosceles, equilateral, and right triangles

VOCABULARY

If an isosceles triangle has exactly two congruent sides, the two angles adjacent to the base are **base angles.**

If an isosceles triangle has exactly two congruent sides, the angle opposite the base is the **vertex angle.**

Theorem 4.6 Base Angles Theorem
If two sides of a triangle are congruent, then the angles opposite them are congruent.

Theorem 4.7 Converse of the Base Angles Theorem
If two angles of a triangle are congruent, then the sides opposite them are congruent.

Corollary to Theorem 4.6
If a triangle is equilateral, then it is equiangular.

Corollary to Theorem 4.7
If a triangle is equiangular, then it is equilateral.

Theorem 4.8 Hypotenuse-Leg (HL) Congruence Theorem
If the hypotenuse and a leg of a right triangle are congruent to the hypotenuse and a leg of a second right triangle, then the two triangles are congruent.

EXAMPLE 1 Using Properties of Right Triangles

Given that $\angle A$ and $\angle D$ are right angles and $\overline{AB} \cong \overline{DC}$, show that $\triangle ABC \cong \triangle DCB$.

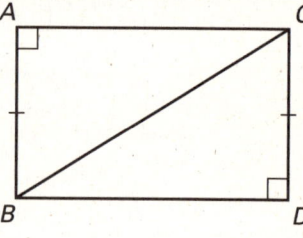

SOLUTION

Paragraph proof   You are given that $\angle A$ and $\angle D$ are right angles. By definition, $\triangle ABC$ and $\triangle DCB$ are right triangles. You are also given that a leg of $\triangle ABC$, $\overline{AB}$, is congruent to a leg of $\triangle DCB$, $\overline{DC}$. You know that the hypotenuses of these two triangles, $\overline{BC}$ for both triangles, are congruent because $\overline{BC} \cong \overline{BC}$ by the Reflexive Property of Congruence. Thus, by the Hypotenuse-Leg Congruence Theorem, $\triangle ABC \cong \triangle DCB$.

LESSON 4.6 CONTINUED

Practice with Examples
For use with pages 236–242

Exercises for Example 1

Write a paragraph proof.

1. Given: $\overline{BC} \perp \overline{AD}$, $\overline{AB} \cong \overline{DB}$
   Prove: $\triangle ABC \cong \triangle DBC$

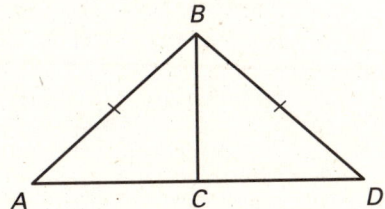

2. Given: $m\angle JKL = m\angle MLK = 90°$,
   $\overline{JL} \cong \overline{MK}$
   Prove: $\overline{JK} \cong \overline{ML}$

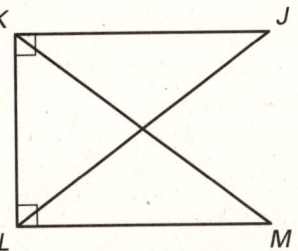

EXAMPLE 2   Using Equilateral and Isosceles Triangles

Find the values of x and y.

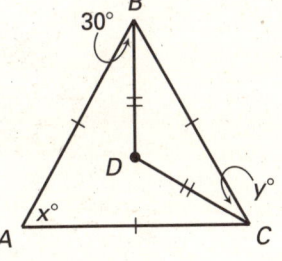

SOLUTION

Notice that $\triangle ABC$ is an equilateral triangle. By the Corollary to Theorem 4.6, $\triangle ABC$ is also an equiangular triangle. Thus $m\angle A = m\angle ABC = m\angle ACB = 60°$. So, $x = 60$.

Notice also that $\triangle DBC$ is an isosceles triangle, and thus by the Base Angles Theorem, $m\angle DBC = m\angle DCB$. Now, since $m\angle ABC = m\angle ABD + m\angle DBC$, $m\angle DBC = 60° - 30° = 30°$. Thus, $y = 30$ by substitution.

LESSON 4.6 CONTINUED

Practice with Examples
For use with pages 236–242

Exercises for Example 2
Find the values of x and y.

3.
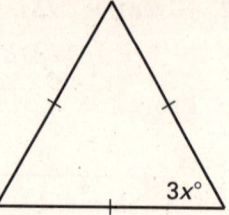

4.
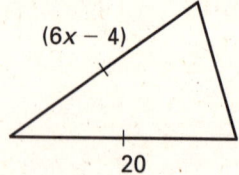

5.
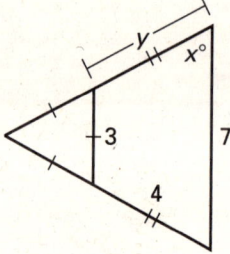

LESSON 4.7

NAME _____ DATE _____

Practice with Examples

For use with pages 243–250

GOAL Place geometric figures in a coordinate plane and write a coordinate proof

VOCABULARY

A **coordinate proof** involves placing geometric figures in a coordinate plane and then using the Distance Formula and the Midpoint Formula, as well as postulates and theorems, to prove statements about the figures.

EXAMPLE 1 Using the Distance Formula

A right triangle has legs of 6 units and 8 units. Place the triangle in a coordinate plane. Label the coordinates of the vertices and find the length of the hypotenuse.

SOLUTION

One possible placement of the triangle is shown. Points on the same vertical segment have the same x-coordinate, and points on the same horizontal segment have the same y-coordinate.

Use the Distance Formula to find d.

$d = \sqrt{(x_2 - x_1)^2 + (y_2 - y_1)^2}$ Distance Formula

$= \sqrt{(8 - 0)^2 + (6 - 0)^2}$ Substitute.

$= \sqrt{100} = 10$ Simplify and evaluate square root.

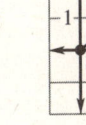

Exercises for Example 1

Use a coordinate plane and the Distance Formula to answer the question.

1. A rectangle has sides of length 8 units and 2 units. What is the length of one of the diagonals?

LESSON 4.7 CONTINUED

Practice with Examples
For use with pages 243–250

EXAMPLE 2 Using the Midpoint Formula

In the diagram, $\triangle ABD \cong \triangle CBD$.

Find the coordinates of point B.

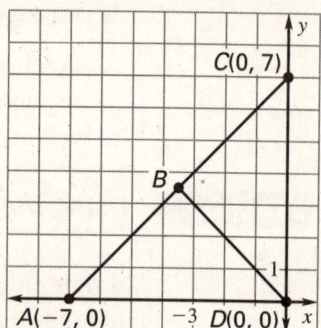

SOLUTION

Because the triangles are congruent, $\overline{AB} \cong \overline{CB}$. So, point B is the midpoint of $\overline{AC}$. This means you can use the Midpoint Formula to find the coordinates of point B.

$B(x, y) = \left(\dfrac{x_1 + x_2}{2}, \dfrac{y_1 + y_2}{2}\right)$ Midpoint Formula

$= \left(\dfrac{-7 + 0}{2}, \dfrac{0 + 7}{2}\right) = \left(-\dfrac{7}{2}, \dfrac{7}{2}\right)$ Substitute and Simplify.

Exercises for Example 2

2. In the diagram, $\triangle FGH \cong \triangle JIH$. Find the coordinates of H.

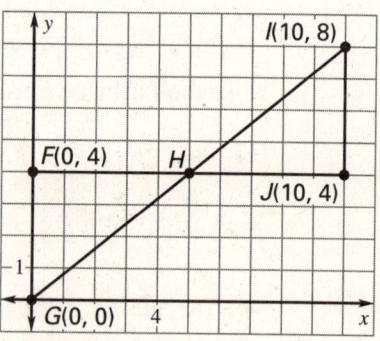

LESSON 4.7 CONTINUED

Practice with Examples
For use with pages 243–250

EXAMPLE 3 Writing a Coordinate Proof

Write a plan to prove that △DEF ≅ △DGF.

Given: Coordinates of figure DEFG.

Prove: △DEF ≅ △DGF

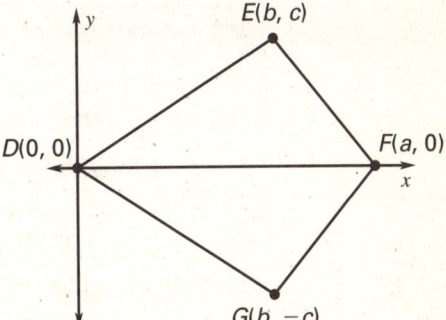

SOLUTION

Plan for Proof Use the Distance Formula to show that segments $\overline{EF}$ and $\overline{GF}$ have equal lengths and that segments $\overline{DE}$ and $\overline{DG}$ have equal lengths. Use the Reflexive Property of Congruence to show that $\overline{DF} \cong \overline{DF}$. Then use SSS Congruence Postulate to conclude that △DEF ≅ △DGF.

Exercises for Example 3

Describe a plan for the proof.

3. **Given:** Coordinates of figure ABCD.

   Prove: △ABC ≅ △CDA

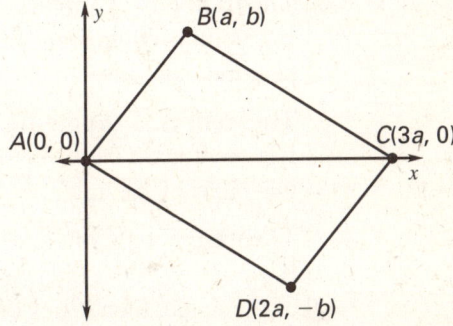

LESSON 5.1

Practice with Examples
For use with pages 264–271

GOAL Use properties of perpendicular bisectors and use properties of angle bisectors to identify equal distances

VOCABULARY

A segment, ray, line, or plane that is perpendicular to a segment at its midpoint is called a **perpendicular bisector.**

A point is **equidistant from two points** if its distance from each point is the same.

The **distance from a point to a line** is defined as the length of the perpendicular segment from the point to the line.

When a point is the same distance from a line as it is from another line, then the point is **equidistant from the two lines** (or rays or segments).

Theorem 5.1 Perpendicular Bisector Theorem
If a point is on the perpendicular bisector of a segment, then it is equidistant from the endpoints of the segment.

Theorem 5.2 Converse of the Perpendicular Bisector Theorem
If a point is equidistant from the endpoints of a segment, then it is on the perpendicular bisector of the segment.

Theorem 5.3 Angle Bisector Theorem
If a point is on the bisector of an angle, then it is equidistant from the two sides of the angle.

Theorem 5.4 Converse of the Angle Bisector Theorem
If a point is in the interior of an angle and is equidistant from the sides of the angle, then it lies on the bisector of the angle.

EXAMPLE 1 *Using Perpendicular Bisectors*

In the diagram shown, $\overrightarrow{EC}$ is the perpendicular bisector of $\overline{AB}$ and $\overline{AF} \cong \overline{BF}$.

  a. Explain how you know that $AC = BC$.
  b. Explain why F is on $\overrightarrow{EC}$.

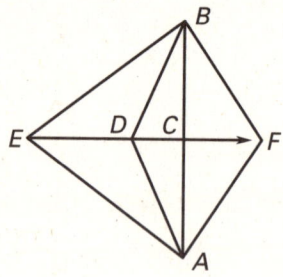

SOLUTION

  a. $\overrightarrow{EC}$ bisects $\overline{AB}$, so $AC = BC$ by the definition of bisector.

  b. $\overline{AF} \cong \overline{BF}$ and by definition of congruence, this means that $AF = BF$ and hence F is equidistant from A and B. By Theorem 5.2, F is on the perpendicular bisector of $\overline{AB}$, which is $\overrightarrow{EC}$.

LESSON 5.1 CONTINUED

Practice with Examples
For use with pages 264–271

Exercises for Example 1

Use the diagram shown. In the diagram, $\overleftrightarrow{AB}$ is the perpendicular bisector of $\overline{CD}$.

1. Find the value of x.

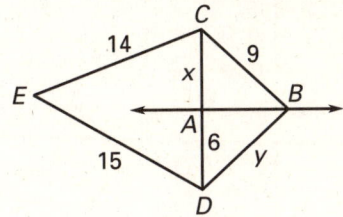

2. Find the value of y.

3. Is E on $\overleftrightarrow{AB}$? Explain.

EXAMPLE 2 Using Bisector Theorems

Determine the correct measurement for the angle or segment given.

a. $\angle DCB$
b. $\overline{FE}$
c. $\overline{AC}$

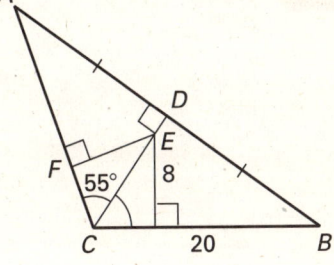

SOLUTION

a. $\overline{CD}$ is the angle bisector of $\angle ACB$ because $m\angle ACD = m\angle DCB$. Since you are given that $m\angle ACD = 55°$, $m\angle DCB = 55°$.

b. By Theorem 5.3, E is equidistant from $\overline{AC}$ and $\overline{BC}$. So $FE = 8$.

c. Because $\overline{CD}$ is the perpendicular bisector of $\overline{AB}$, then by Theorem 5.1 C is equidistant from A and B. Thus, $AC = 20$.

LESSON 5.1 CONTINUED

Practice with Examples
For use with pages 264–271

Exercises for Example 2

Determine the correct measurement for the angle or segment given.

4. $\overline{EG}$

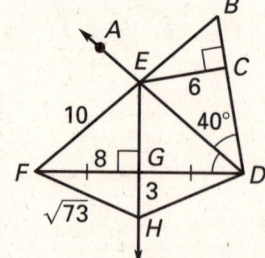

5. $\angle GDE$

6. $\overline{ED}$

7. $\overline{HD}$

8. $\overline{FD}$

LESSON 5.2

NAME _____ DATE _____

Practice with Examples
For use with pages 272–278

GOAL Use properties of perpendicular bisectors of a triangle and use properties of angle bisectors of a triangle

VOCABULARY

A **perpendicular bisector** of a triangle is a line (or ray or segment) that is perpendicular to a side of the triangle at the midpoint of the side.

When three or more lines (or rays or segments) intersect in the same point, they are called **concurrent lines** (or rays or segments).

The point of intersection of concurrent lines is called the **point of concurrency.**

The point of concurrency of the perpendicular bisectors of a triangle is called the **circumcenter** of the triangle.

An **angle bisector** of a triangle is a bisector of an angle of the triangle.

The point of concurrency of the angle bisectors is called the **incenter** of the triangle.

Theorem 5.5 Concurrency of Perpendicular Bisectors of a Triangle
The perpendicular bisectors of a triangle intersect at a point that is equidistant from the vertices of the triangle.

Theorem 5.6 Concurrency of Angle Bisectors of a Triangle
The angle bisectors of a triangle intersect at a point that is equidistant from the sides of the triangle.

EXAMPLE 1 Using Perpendicular Bisectors

The perpendicular bisectors of △ABC meet at point D.

    a. Find DB.

    b. Find AE.

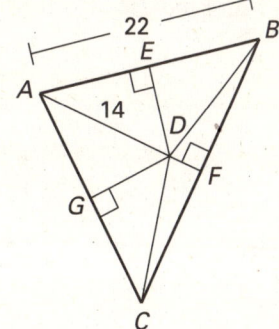

SOLUTION

    a. You are given that the perpendicular bisectors of △ABC meet at D. By Theorem 5.5, D is equidistant from the vertices A, B, and C of the triangle. Since you are given that AD = 14, it follows that DB = 14.

    b. You are given that $\overline{ED}$ is a perpendicular bisector of side $\overline{AB}$. By definition of a perpendicular bisector, E is the midpoint of $\overline{AB}$. Because you are given that AB = 22, it follows that AE = 11.

LESSON 5.2 CONTINUED

Practice with Examples
For use with pages 272–278

Exercises for Example 1

Use the given information to find the indicated lengths.

1. The perpendicular bisectors of △HIJ meet at K, IJ = 18, and KJ = 12.

   a. Find HK.

   b. Find IM.

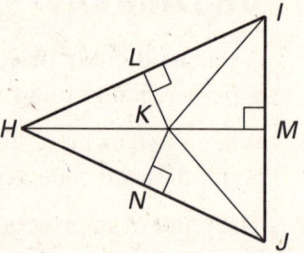

2. R is the circumcenter of △OPQ, OS = 10, QR = 12, and PQ = 22.

   a. Find OP.

   b. Find RP.

   c. Find OR.

   d. Find TP.

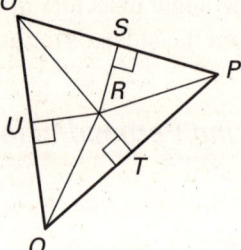

LESSON 5.2 CONTINUED

Practice with Examples
For use with pages 272–278

EXAMPLE 2 Using Angle Bisectors

The angle bisectors of △ABC meet at point D. Find DE.

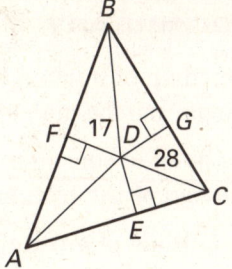

SOLUTION

By Theorem 5.6, point D is equidistant from the sides of △ABC. Thus, FD = GD = ED. Since FD = 17 units and FD = ED, it follows that DE = 17.

Exercise for Example 2

3. The angle bisectors of △ABC meet at point P, PR = 3, and PC = 5. Find QP.

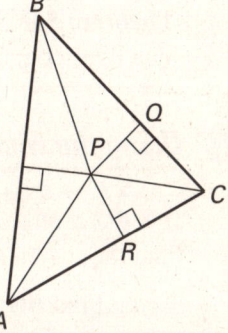

LESSON 5.3

NAME _____ DATE _____

Practice with Examples

For use with pages 279–285

GOAL Use properties of medians of a triangle and use properties of altitudes of a triangle

VOCABULARY

A **median of a triangle** is a segment whose endpoints are a vertex of the triangle and the midpoint of the opposite side.

The point of concurrency of the three medians of a triangle is called the **centroid of the triangle.**

An **altitude of a triangle** is the perpendicular segment from a vertex to the opposite side or to the line that contains the opposite side.

The lines containing the three altitudes are concurrent and intersect at a point called the **orthocenter of the triangle.**

Theorem 5.7 Concurrency of Medians of a Triangle
The medians of a triangle are concurrent at a point that is two thirds of the distance from each vertex to the midpoint of the opposite side.

Theorem 5.8 Concurrency of Altitudes of a Triangle
The lines containing the altitudes of a triangle are concurrent.

EXAMPLE 1 Using the Medians of a Triangle

D is the centroid of $\triangle ABC$ and $DG = 4$.
Find the indicated values.

   a. Find BG.

   b. Find BD.

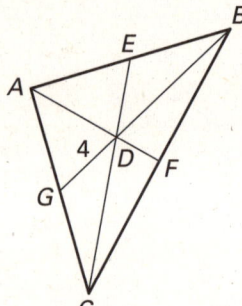

SOLUTION

a. Because D is the centroid of $\triangle ABC$, $BD = \frac{2}{3}BG$. Then
$DG = BG - BD = \frac{1}{3}BG$. Substituting 4 for DG, $4 = \frac{1}{3}BG$, so $BG = 12$.

b. $BD = \frac{2}{3}BG$, so by substituting 12 for BG, you get $BD = \frac{2}{3}(12) = 8$. So, $BD = 8$.

LESSON 5.3 CONTINUED

Practice with Examples

For use with pages 279–285

Exercises for Example 1

Use the figure and the given information. *D* is the centroid of $\triangle ABC$, $\overline{BE} \perp \overline{AC}$, $\overline{AB} \cong \overline{CB}$, $FB = 5$, $EC = 3$, and $DF = 2$.

1. Find *CF*.
2. Find *CG*.

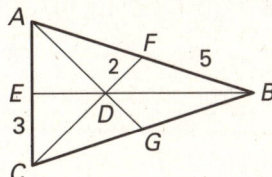

3. Find *CD*.

4. Find the perimeter of $\triangle ABC$.

EXAMPLE 2 *Finding the Centroid of a Triangle*

Find the coordinates of the centroid of $\triangle ABC$.

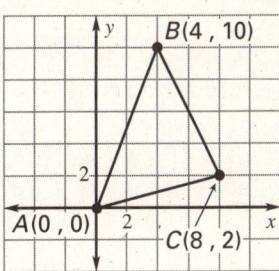

SOLUTION

You know that the centroid is two thirds of the distance from each vertex to the midpoint of the opposite side. If you find the midpoint of any side and draw a segment from that point to the opposite vertex, you will have a segment which contains the centroid. You can then find the length of the segment. Finally, you know that the centroid is two thirds of the length of this segment from the vertex.

Find the midpoint of $\overline{AC}$. The midpoint of $\overline{AC}$ is $\left(\dfrac{0+8}{2}, \dfrac{0+2}{2}\right) = (4, 1)$. A median can be drawn from this midpoint to the vertex *B*. Use the distance formula to find the length of this median.

$$d = \sqrt{(4-4)^2 + (10-1)^2} = \sqrt{0^2 + 9^2} = 9$$

By Theorem 5.7, the centroid is two thirds of this distance down from *B* along the median $\left(\dfrac{2}{3} \cdot 9 = 6\right)$. The coordinates of the centroid are $(4, 10-6)$, or $(4, 4)$.

Geometry
Practice Workbook with Examples

LESSON 5.3 CONTINUED

Practice with Examples

For use with pages 279–285

Exercises for Example 2

Find the coordinates of the centroid of the triangle with the given vertices.

5. $A(0, 0)$, $B(10, 0)$, $C(5, 6)$

6. $D(-5, 2)$, $E(-3, 6)$, $F(-7, 10)$

7. $G(-1, 2)$, $H(-3, 10)$, $I(10, 6)$ (*Hint:* Find the median from I to $\overline{GH}$.)

LESSON 5.4

Practice with Examples
For use with pages 287–293

GOAL Identify the midsegments of a triangle and use properties of midsegments of a triangle

VOCABULARY

A **midsegment of a triangle** is a segment that connects the midpoints of two sides of a triangle.

Theorem 5.9 Midsegment Theorem
The segment connecting the midpoints of two sides of a triangle is parallel to the third side and is half as long.

EXAMPLE 1 Using the Midsegment Theorem

Show that the midsegment $\overline{ED}$ is parallel to side $\overline{BC}$ and is half as long.

SOLUTION

Use the Midpoint Formula to find the coordinates of D and E.

$$D = \left(\frac{-3+7}{2}, \frac{0+(-2)}{2}\right) = (2, -1)$$

$$E = \left(\frac{-3+3}{2}, \frac{0+6}{2}\right) = (0, 3)$$

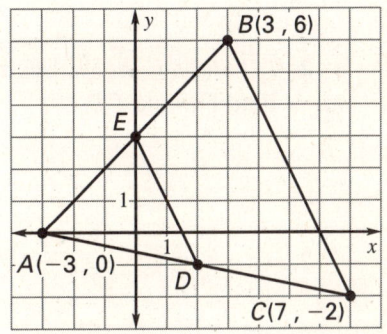

Next, find the slopes of $\overline{BC}$ and $\overline{ED}$.

Slope of $\overline{BC} = \dfrac{6-(-2)}{3-7} = -\dfrac{8}{4} = -2$ Slope of $\overline{ED} = \dfrac{3-(-1)}{0-2} = -\dfrac{4}{2} = -2$

Because their slopes are equal, $\overline{BC}$ and $\overline{ED}$ are parallel. You can use the Distance Formula to show that $ED = \sqrt{20} = 2\sqrt{5}$ and $BC = \sqrt{80} = 4\sqrt{5}$. So $\overline{ED}$ is half as long as $\overline{BC}$.

LESSON 5.4 CONTINUED

Practice with Examples
For use with pages 287–293

Exercises for Example 1

Use the Midsegment Theorem.

1. Show that the midsegment $\overline{ED}$ is parallel to side $\overline{BC}$ and is half as long.

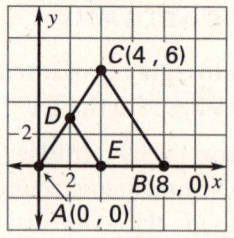

2. $\overline{ED}$ and $\overline{DF}$ are midsegments in $\triangle ABC$. Find DF and CB.

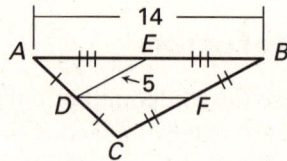

3. Given $PQ = 14$, $SU = 6$, and $QU = 3$, find the perimeter of $\triangle STU$.

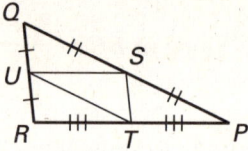

LESSON 5.4 CONTINUED

NAME _____ DATE _____

Practice with Examples
For use with pages 287–293

EXAMPLE 2 Using Midpoints to Draw a Triangle

The midpoints of the sides of a triangle are $A(-4, 1)$, $B(-2, 2)$, and $C(-5, 3)$. What are the coordinates of the vertices of the triangle?

SOLUTION

Plot the midpoints in a coordinate plane.

Connect these midpoints to form the midsegments $\overline{AB}$, $\overline{BC}$, and $\overline{CA}$.

Find the slopes of these midsegments. Use the slope formula as shown.

$$\text{slope} = \frac{3-2}{-5-(-2)} = -\frac{1}{3}$$

$$\text{slope} = \frac{1-3}{-4-(-5)} = -2$$

$$\text{slope} = \frac{2-1}{-2-(-4)} = \frac{1}{2}$$

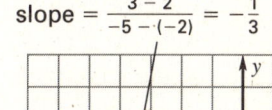

Each midsegment contains two of the unknown triangle's midpoints and is parallel to the side that contains the third midpoint (by the Midsegment Theorem). So, you know a point on each side of the unknown triangle and the slope of each side (because parallel lines have equal slopes).

Draw the lines that contain the three sides.

The lines intersect at $D(-7, 2)$, $E(-3, 4)$, and $F(-1, 0)$, which are the vertices of the triangle.

Exercises for Example 2

You are given the midpoints of the sides of a triangle. Find the coordinates of the vertices of the triangle.

4. $L(0, 3)$, $M(2, -3)$, $N(-4, 5)$

5. $L(-7, 5)$, $M(-1, -1)$, $N(3, 1)$

6. $L(1, 3)$, $M(4, -2)$, $N(7, 1)$

7. $L(6, 3)$, $M(1, 0)$, $N(-2, 4)$

LESSON 5.5

Practice with Examples
For use with pages 295–301

GOAL Compare measurements of a triangle to decide which side is longest or which angle is largest and use the Triangle Inequality

VOCABULARY

Theorem 5.10
If one side of a triangle is longer than another side, then the angle opposite the longer side is larger than the angle opposite the shorter side.

Theorem 5.11
If one angle of a triangle is larger than another angle, then the side opposite the larger angle is longer than the side opposite the smaller angle.

Theorem 5.12   Exterior Angle Inequality
The measure of an exterior angle of a triangle is greater than the measure of either of the two nonadjacent interior angles.

Theorem 5.13   Triangle Inequality
The sum of the lengths of any two sides of a triangle is greater than the length of the third side.

EXAMPLE 1   Writing Measurements in Order from Least to Greatest

Write the measurements of the triangle in order from least to greatest.

a.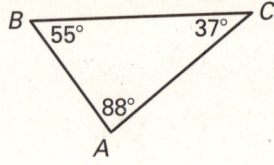

b.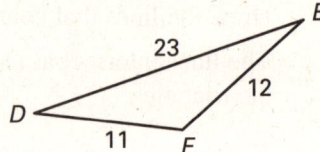

SOLUTION

a. $m\angle C < m\angle B < m\angle A$
   $AB < AC < BC$

b. $DF < EF < DE$
   $m\angle E < m\angle D < m\angle F$

---

94   Geometry
Practice Workbook with Examples

LESSON 5.5 CONTINUED

Practice with Examples

For use with pages 295–301

Exercises for Example 1

Write the measurements of the triangle in order from least to greatest.

1.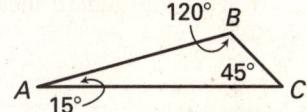

2.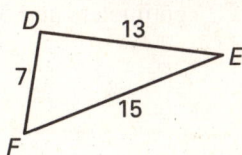

3.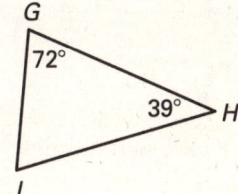

4.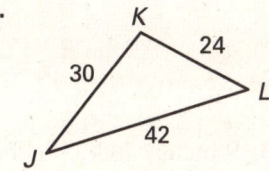

EXAMPLE 2 Finding Possible Side Lengths

A triangle has one side of 12 inches and another side of 16 inches. Describe the possible lengths of the third side.

SOLUTION

Let x represent the length of the third side. Using the Triangle Inequality, you can write and solve inequalities.

$$x + 12 > 16 \qquad 16 + 12 > x$$
$$x > 4 \qquad\qquad 28 > x$$

So, the length of the third side must be greater than 4 inches and less than 28 inches.

LESSON 5.5 CONTINUED

Practice with Examples
For use with pages 295–301

Exercises for Example 2

Two sides of a triangle are given. Describe the possible lengths of the third side.

5. 2 centimeters and 5 centimeters

6. 7 inches and 12 inches

7. 4 feet and 10 feet

8. 11 meters and 10 meters

9. 9 inches and 25 inches

10. 1 mile and 8 miles

LESSON 5.6

Practice with Examples
For use with pages 302–308

GOAL Read and write an indirect proof and use the Hinge Theorem and its converse to compare side lengths and angle measures

VOCABULARY

An **indirect proof** is a proof in which you prove that a statement is true by first assuming that its opposite is true. If this assumption leads to an impossibility, then you have proved that the original statement is true.

Theorem 5.14   Hinge Theorem
If two sides of one triangle are congruent to two sides of another triangle, and the included angle of the first is larger than the included angle of the second, then the third side of the first is longer than the third side of the second.

Theorem 5.15   Converse of the Hinge Theorem
If two sides of one triangle are congruent to two sides of another triangle, and the third side of the first is longer than the third side of the second, then the included angle of the first is larger than the included angle of the second.

EXAMPLE 1   Using the Hinge Theorem

Complete with $<$, $>$, or $=$.

a. BC ____ EF

b. JG ____ JI

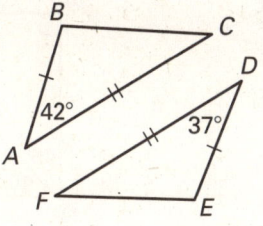
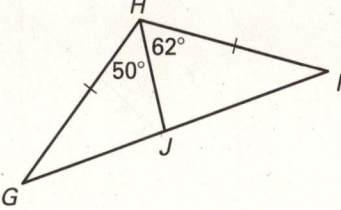

c. $m\angle M$ ____ $m\angle N$

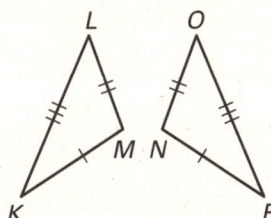

SOLUTION

a. By the Hinge Theorem, since $\overline{AB} \cong \overline{DE}$, $\overline{AC} \cong \overline{DF}$, and $m\angle A > m\angle D$, it follows that $BC > EF$.

b. Since $m\angle JHG < m\angle JHI$, $JG < JI$.

c. $m\angle M = m\angle N$

LESSON 5.6 CONTINUED

Practice with Examples
For use with pages 302–308

Exercises for Example 1

Complete with <, >, or =.

1. $m\angle 1$ ____ $m\angle 2$

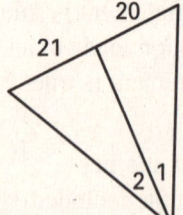

2. BC ____ EF

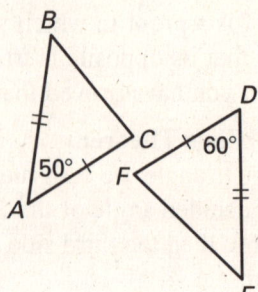

3. GI ____ JL

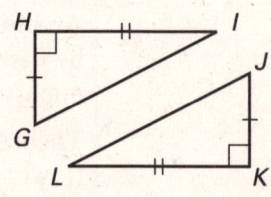

4. MN ____ QP

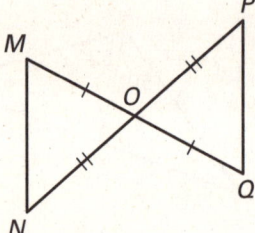

5. $m\angle 1$ ____ $m\angle 2$

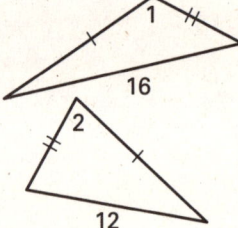

6. AD ____ BC

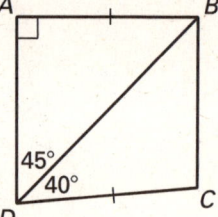

LESSON 5.6 CONTINUED

Practice with Examples
For use with pages 302–308

EXAMPLE 2 Finding Possible Side Lengths and Angle Measures

You can use the Hinge Theorem and its converse to choose possible side lengths or angle measures from a given list.

At the right, $\overline{AB} \cong \overline{DE}$, $\overline{BC} \cong \overline{EF}$, $AC = 10$ cm, $m\angle B = 60°$, and $m\angle E = 130°$. Which of the following is a possible length for $\overline{DF}$: 5 cm, 10 cm, or 15 cm?

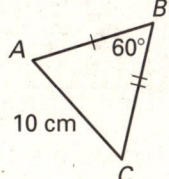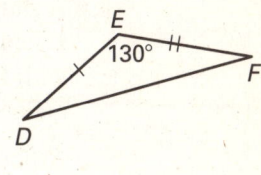

SOLUTION

Because the included angle in $\triangle DEF$ is larger than the included angle in $\triangle ABC$, the third side $\overline{DF}$ must be longer than $\overline{AC}$. So of the three choices, the only possible length for $\overline{DF}$ is 15 cm.

Exercise for Example 2

Choose possible side lengths or angle measures from the given list.

7. In a $\triangle ABC$ and a $\triangle XYZ$, $\overline{AB} \cong \overline{YZ}$, $\overline{AC} \cong \overline{YX}$, $BC = 9$ inches, $XZ = 11$ inches, and $m\angle A = 47°$. Which of the following is a possible measure for $\angle Y$: 35°, 42°, 47°, or 53°?

LESSON 6.1

Practice with Examples
For use with pages 322–328

GOAL Identify, name, and describe polygons and use the sum of the measures of the interior angles of a quadrilateral

> ### VOCABULARY
>
> A **polygon** is a plane figure that is formed by three or more segments called sides, such that no two sides with a common endpoint are collinear, and each side intersects exactly two other sides, one at each endpoint. Each endpoint of a side is a **vertex** of the polygon.
>
> A polygon is **convex** if no line that contains a side of the polygon contains a point in the interior of the polygon.
>
> A polygon that is not convex is called **nonconvex** or **concave.**
>
> A **diagonal** of a polygon is a segment that joins two *nonconsecutive* vertices.
>
> **Theorem 6.1   Interior Angles of a Quadrilateral**
> The sum of the measures of the interior angles of a quadrilateral is 360°.

EXAMPLE 1 Identifying Polygons

State whether the figure is a polygon. If it is not, explain why.

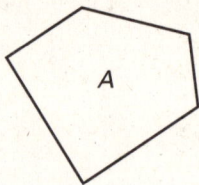 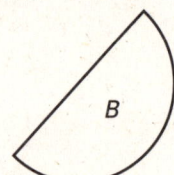 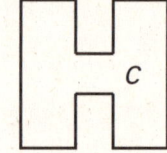 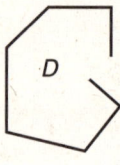

SOLUTION

Figures *A* and *C* are polygons.

- Figure *B is not* a polygon because it only has two sides, and one of its sides is not a segment.

- Figure *D is not* a polygon because two of the sides intersect only one other side.

LESSON 6.1 CONTINUED

Practice with Examples

For use with pages 322–328

Exercises for Example 1

State whether each figure is a polygon. If it is not, explain why.

1.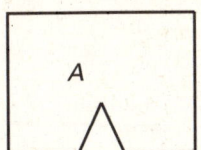
2.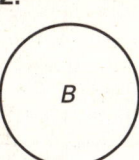
3.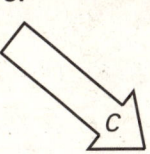
4.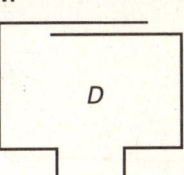

EXAMPLE 2 — Identifying Convex and Concave Polygons

State whether each polygon is convex or concave.

a.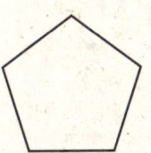

b.

SOLUTION

a. The polygon has 5 sides. When extended, none of the sides intersect the interior, so the polygon is convex.

b. The polygon has 10 sides. When extended, some of the sides intersect the interior, so the polygon is concave.

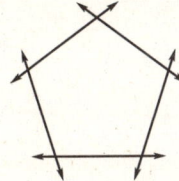
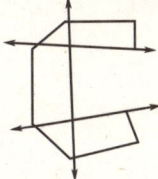

Exercises for Example 2

State whether the polygon is *convex* or *concave*.

5.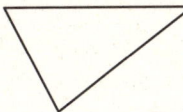
6.
7.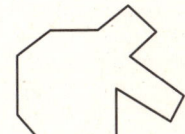

LESSON 6.1 CONTINUED

NAME _____ DATE _____

Practice with Examples

For use with pages 322–328

EXAMPLE 3 Interior Angles of a Quadrilateral

Find $m\angle A$ and $m\angle B$.

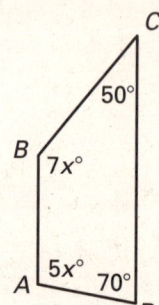

SOLUTION

Find the value of x. Use the Interior Angles of a Quadrilateral Theorem to write an equation involving x. Then solve the equation.

$5x° + 7x° + 50° + 70° = 360°$ Theorem 6.1

$x = 20$ Solve for x.

So, $m\angle A = 5x° = 5(20)° = 100°$ and $m\angle B = 7x° = 7(20)° = 140°$.

Exercises for Example 3

Use the information in the diagram to solve for x.

8.
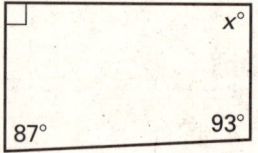

9.
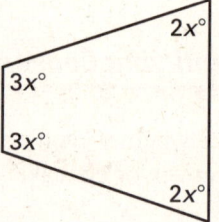

102 Geometry
Practice Workbook with Examples

LESSON 6.2

Practice with Examples

For use with pages 330–337

GOAL Use some properties of parallelograms

VOCABULARY

A **parallelogram** is a quadrilateral with both pairs of opposite sides parallel.

Theorem 6.2
If a quadrilateral is a parallelogram, then its opposite sides are congruent.

Theorem 6.3
If a quadrilateral is a parallelogram, then its opposite angles are congruent.

Theorem 6.4
If a quadrilateral is a parallelogram, then its consecutive angles are supplementary.

Theorem 6.5
If a quadrilateral is a parallelogram, then its diagonals bisect each other.

EXAMPLE 1 Using Properties of Parallelograms

ABCD is a parallelogram. Find the lengths and angle measures.

a. *AD*
b. *EC*
c. $m\angle ADC$
d. $m\angle BCD$

SOLUTION

a. $AD = BC$ from Theorem 6.2. So, $AD = 8$.

b. From Theorem 6.5, the two diagonals of *ABCD* bisect each other. Therefore, $AE = EC$. So, $EC = 5$.

c. $m\angle ABC = m\angle ADC$ from Theorem 6.3.
$m\angle ABC = m\angle ABE + m\angle CBE$ by the Angle Addition Postulate.
Substituting, $m\angle ADC = 65° + 45° = 110°$.

d. $m\angle BCD + m\angle ADC = 180°$ by Theorem 6.4. So $m\angle BCD = 180° - m\angle ADC$ by the Subtraction Property of Equality. By substituting and simplifying, $m\angle BCD = 70°$.

LESSON 6.2 CONTINUED

Practice with Examples

For use with pages 330–337

Exercises for Example 1

Find the value of each variable in the parallelogram.

1.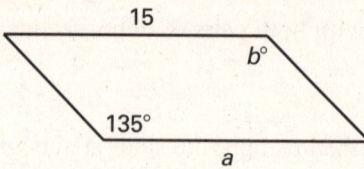

2.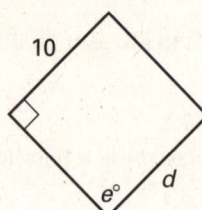

3.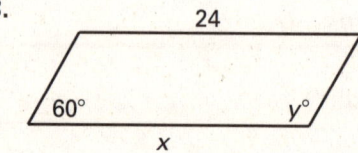

4.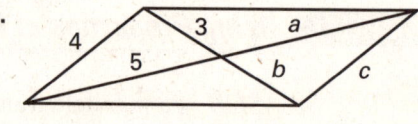

5.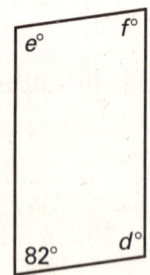

6.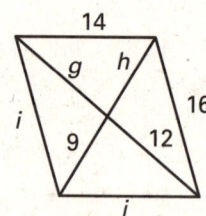

LESSON 6.2 CONTINUED

Practice with Examples
For use with pages 330–337

EXAMPLE 2 Using Algebra with Parallelograms

Use algebra to find the value of each variable in the parallelogram.

a.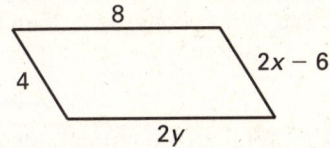

b.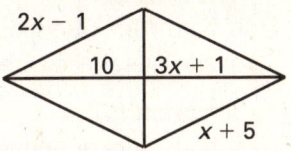

SOLUTION

Set up equations based upon the properties of parallelograms provided in Theorems 6.2 through 6.5.

a. Because opposite sides of a parallelogram are congruent (Theorem 6.2), $2x - 6 = 4$. Solving for x yields $2x = 10$ which means $x = 5$. Also by Theorem 6.2, $2y = 8$, so $y = 4$.

b. From Theorem 6.2, $2x - 1 = x + 5$. Thus, $x = 6$.
From Theorem 6.5, $3x + 1 = 10$. Thus, $3x = 9$ which means $x = 3$.

Exercises for Example 2

Find the value of each variable in the parallelogram.

7.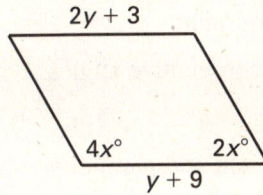

8.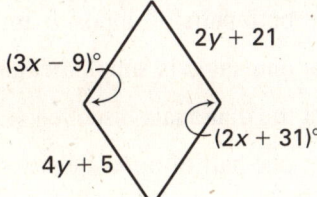

9.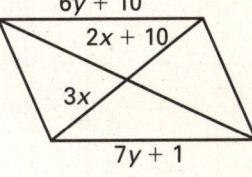

LESSON 6.3

Practice with Examples

For use with pages 338–346

GOAL Prove that a quadrilateral is a parallelogram and use coordinate geometry with parallelograms

---

Theorem 6.6
If both pairs of opposite sides of a quadrilateral are congruent, then the quadrilateral is a parallelogram.

Theorem 6.7
If both pairs of opposite angles of a quadrilateral are congruent, then the quadrilateral is a parallelogram.

Theorem 6.8
If an angle of a quadrilateral is supplementary to both of its consecutive angles, then the quadrilateral is a parallelogram.

Theorem 6.9
If the diagonals of a quadrilateral bisect each other, then the quadrilateral is a parallelogram.

Theorem 6.10
If one pair of opposite sides of a quadrilateral are congruent and parallel, then the quadrilateral is a parallelogram.

Ways to Prove a Shape is a Parallelogram

- Show that both pairs of opposite sides are parallel.
- Show that both pairs of opposite sides are congruent.
- Show that both pairs of opposite angles are congruent.
- Show that one angle is supplementary to both consecutive angles.
- Show that the diagonals bisect each other.
- Show that one pair of opposite sides are congruent and parallel.

LESSON 6.3 CONTINUED

Practice with Examples
For use with pages 338–346

EXAMPLE 1 Using Properties of Parallelograms

Show that $A(2, 0)$, $B(3, 4)$, $C(-2, 6)$, and $D(-3, 2)$ are the vertices of a parallelogram.

SOLUTION

There are many ways to solve this problem.

Method 1 Show that opposite sides have the same slope, so they are parallel.

Slope of $\overline{AB} = \dfrac{4 - 0}{3 - 2} = 4$

Slope of $\overline{CD} = \dfrac{2 - 6}{-3 - (-2)} = \dfrac{-4}{-1} = 4$

Slope of $\overline{BC} = \dfrac{6 - 4}{-2 - 3} = \dfrac{2}{-5} = -\dfrac{2}{5}$

Slope of $\overline{DA} = \dfrac{0 - 2}{2 - (-3)} = \dfrac{-2}{5} = -\dfrac{2}{5}$

$\overline{AB}$ and $\overline{CD}$ have the same slope, so they are parallel. Similarly, $\overline{BC} \parallel \overline{DA}$. Because opposite sides are parallel, $ABCD$ is a parallelogram.

Method 2 Show that the opposite sides have the same length.

$AB = \sqrt{(3 - 2)^2 + (4 - 0)^2} = \sqrt{17}$

$CD = \sqrt{(-3 - (-2))^2 + (2 - 6)^2} = \sqrt{17}$

$BC = \sqrt{(-2 - 3)^2 + (6 - 4)^2} = \sqrt{29}$

$DA = \sqrt{(2 - (-3))^2 + (0 - 2)^2} = \sqrt{29}$

$\overline{AB} \cong \overline{CD}$ and $\overline{BC} \cong \overline{DA}$. Because both pairs of opposite sides are congruent, $ABCD$ is a parallelogram.

Method 3 Show that one pair of opposite sides is congruent and parallel. Find the slopes and lengths of $\overline{AB}$ and $\overline{CD}$ as shown in Methods 1 and 2.

Slope of $\overline{AB} =$ Slope of $\overline{CD} = 4$

$AB = CD = \sqrt{17}$

$\overline{AB}$ and $\overline{CD}$ are congruent and parallel, so $ABCD$ is a parallelogram.

LESSON 6.3 CONTINUED

Practice with Examples

For use with pages 338–346

Exercises for Example 1

Refer to the methods demonstrated in Example 1 to show that the quadrilateral with the given vertices is a parallelogram.

1. Show that the quadrilateral with vertices $A(-3, 0)$, $B(-2, -4)$, $C(-7, -6)$, and $D(-8, -2)$ is a parallelogram using Method 1 from Example 1.

2. Show that the quadrilateral with vertices $A(-4, 1)$, $B(1, 2)$, $C(4, -4)$, and $D(-1, -5)$ is a parallelogram using Method 2 from Example 1.

3. Show that the quadrilateral with vertices $A(0, -6)$, $B(4, -5)$, $C(6, 3)$, and $D(2, 2)$ is a parallelogram using Method 3 from Example 1.

4. Show that the quadrilateral with vertices $A(-1, -2)$, $B(5, -3)$, $C(6, 6)$, and $D(0, 7)$ is a parallelogram using any of the three methods demonstrated in Example 1.

LESSON 6.4

Practice with Examples
For use with pages 347–355

GOAL Use properties of sides and angles of rhombuses, rectangles, and squares and use properties of diagonals of rhombuses, rectangles, and squares

VOCABULARY

A **rhombus** is a parallelogram with four congruent sides.

A **rectangle** is a parallelogram with four right angles.

A **square** is a parallelogram with four congruent sides and four right angles.

Rhombus Corollary
A quadrilateral is a rhombus if and only if it has four congruent sides.

Rectangle Corollary
A quadrilateral is a rectangle if and only if it has four right angles.

Square Corollary
A quadrilateral is a square if and only if it is a rhombus and a rectangle.

Theorem 6.11
A parallelogram is a rhombus if and only if its diagonals are perpendicular.

Theorem 6.12
A parallelogram is a rhombus if and only if each diagonal bisects a pair of opposite angles.

Theorem 6.13
A parallelogram is a rectangle if and only if its diagonals are congruent.

EXAMPLE 1 Using Properties of Special Parallelograms

ABCD is a square. What else do you know about *ABCD*?

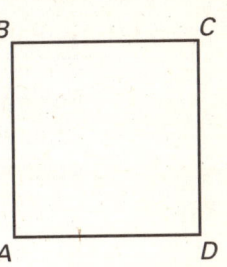

SOLUTION

By the definition of a square, *ABCD* has four right angles and four congruent sides. Also, because squares are parallelograms, *ABCD* has all of the properties of a parallelogram:

- Opposite sides are parallel and congruent.
- Opposite angles are congruent and consecutive angles are supplementary.
- Diagonals bisect each other.

LESSON 6.4 CONTINUED

Practice with Examples
For use with pages 347–355

Exercises for Example 1

State all that you know about the special parallelogram given.

1. ABCD is a rhombus. What else do you know about ABCD?

2. EFGH is a rectangle. What else do you know about EFGH?

EXAMPLE 2 Using Properties of Special Parallelograms

For any rectangle ABCD, decide whether the statement is *always*, *sometimes*, or *never true*.

a. $\angle A \cong \angle C$ b. $\overline{AB} \cong \overline{CD}$ c. $\overline{AB} \cong \overline{BC}$

SOLUTION

a. Always true. By definition, a rectangle has four right angles, so $\angle A$ and $\angle C$ are both right angles, and therefore they are congruent.

b. Always true. $\overline{AB}$ and $\overline{CD}$ are opposite sides. Because all rectangles are parallelograms, opposite sides of a rectangle are congruent.

c. Sometimes true. $\overline{AB}$ and $\overline{BC}$ are adjacent sides. Adjacent sides of a rectangle do not need to be congruent.

Exercises for Example 2

For any rectangle ABCD, decide whether the statement is *always, sometimes*, or *never* true. Draw a sketch and explain your answer.

3. $\angle A \cong \angle C$ 4. $\overline{AB} \cong \overline{BC}$ 5. $\overline{AC} \cong \overline{BD}$

LESSON 6.4 CONTINUED

Practice with Examples
For use with pages 347–355

EXAMPLE 3 *Using Properties of a Rhombus*

In the diagram, *ABCD* is a rhombus.

What is the value of *x*?

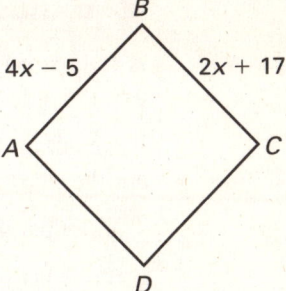

SOLUTION

Because *ABCD* is a rhombus, all four sides are congruent. So, $AB = BC$.

$4x - 5 = 2x + 17$ Equate lengths of congruent sides.

$x = 11$ Solve for *x*.

Exercises for Example 3

Use properties of special quadrilaterals to find the value of *x*.

6. *ABCD* is a rhombus.

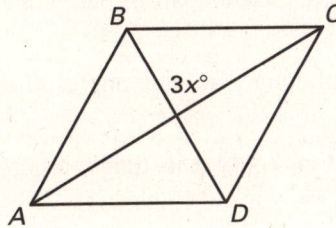

7. *EFGH* is a rectangle.

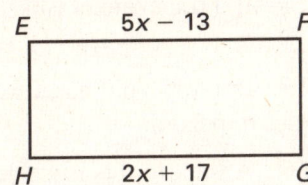

8. *ABCD* is a square.

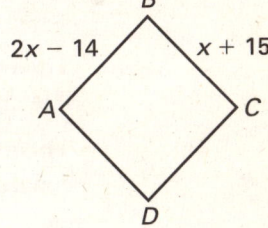

LESSON 6.5

Practice with Examples
For use with pages 356–363

GOAL Use properties of trapezoids and kites

VOCABULARY

A **trapezoid** is a quadrilateral with exactly one pair of parallel sides. The parallel sides of a trapezoid are the **bases** of the trapezoid.

For each of the bases of a trapezoid, there is a pair of **base angles,** which are the two angles that have that base as a side.

The nonparallel sides of a trapezoid are the **legs** of the trapezoid.

If the legs of a trapezoid are congruent, then the trapezoid is an **isosceles trapezoid.**

The **midsegment** of a trapezoid is the segment that connects the midpoints of its legs.

A **kite** is a quadrilateral that has two pairs of consecutive congruent sides, but opposite sides are not congruent.

Theorem 6.14 If a trapezoid is isosceles, then each pair of base angles is congruent.

Theorem 6.15 If a trapezoid has a pair of congruent base angles, then it is an isosceles trapezoid.

Theorem 6.16 A trapezoid is isosceles if and only if its diagonals are congruent.

Theorem 6.17 The midsegment of a trapezoid is parallel to each base and its length is one half the sum of the lengths of its bases.

Theorem 6.18 If a quadrilateral is a kite, then its diagonals are perpendicular.

Theorem 6.19 If a quadrilateral is a kite, then exactly one pair of opposite angles is congruent.

EXAMPLE 1 Finding Midsegment Lengths of Trapezoids and Using Algebra

a. Find the length of the midsegment $\overline{MN}$.

b. Find the value of x.

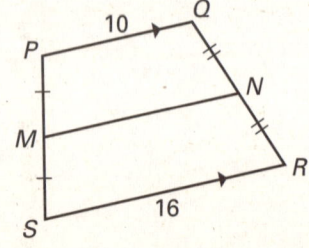
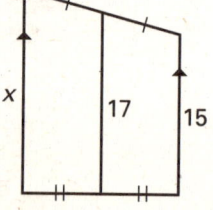

LESSON 6.5 CONTINUED

NAME _____ DATE _____

Practice with Examples
For use with pages 356–363

SOLUTION

a. Use the Midsegment Theorem for Trapezoids.

$$MN = \tfrac{1}{2}(PQ + SR) = \tfrac{1}{2}(10 + 16) = \tfrac{1}{2}(26) = 13$$

b. $17 = \tfrac{1}{2}(15 + x)$ Midsegment Theorem for Trapezoids
   $34 = 15 + x$ Multiply each side by 2.
   $19 = x$ Subtract.

Exercises for Example 1

Find the value of x.

1.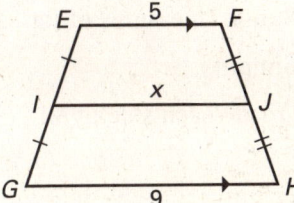

2.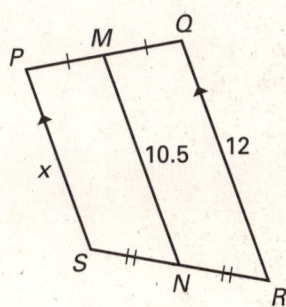

3.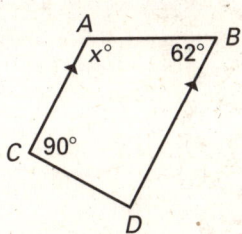

LESSON 6.5 CONTINUED

Practice with Examples
For use with pages 356–363

EXAMPLE 2 — Using Properties of Kites

JKLM is a kite. What is $m\angle J$?

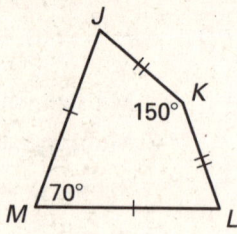

SOLUTION

JKLM is a kite, so $\angle J \cong \angle L$ and $m\angle J = m\angle L$.

$2(m\angle J) + 150° + 70° = 360°$ Sum of measures of int. ∠s of a quad. is 360°.

$2(m\angle J) = 140°$ Simplify.

$m\angle J = 70°$ Divide each side by 2.

Exercises for Example 2

Find the value of *x*.

4.
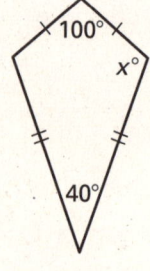

5.
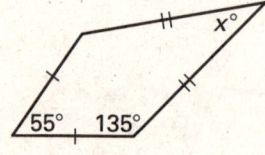

6.
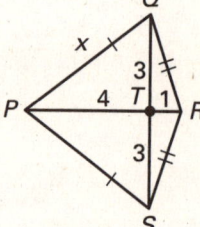

LESSON 6.6

NAME _____ DATE _____

Practice with Examples

For use with pages 364–370

GOAL Identify special quadrilaterals based on limited information and prove that a quadrilateral is a special type of quadrilateral, such as a rhombus or a trapezoid

> **Ways to Prove a Shape is a Rhombus**
>
> 1. You can use the definition and show that the quadrilateral is a *parallelogram* that has four congruent sides. It is easier, however, to use the Rhombus Corollary and simply show that all four sides of the quadrilateral are congruent.
>
> 2. Show that the quadrilateral is a parallelogram and that the diagonals are perpendicular. *(Theorem 6.11)*
>
> 3. Show that the quadrilateral is a parallelogram and that each diagonal bisects a pair of opposite angles. *(Theorem 6.12)*

EXAMPLE 1 Proving a Quadrilateral is a Rhombus

Show that *ABCD* is a rhombus.

SOLUTION

There are several ways of solving this problem.

Method 1 Use the Rhombus Corollary and show that all four sides of the quadrilateral are congruent.

$AB = \sqrt{(1-(-3))^2 + (3-1)^2} = \sqrt{20}$

$BC = \sqrt{(5-1)^2 + (1-3)^2} = \sqrt{20}$

$CD = \sqrt{(1-5)^2 + (-1-1)^2} = \sqrt{20}$

$DA = \sqrt{(-3-1)^2 + (1-(-1))^2} = \sqrt{20}$

So, because $AB = BC = CD = DA$, *ABCD* is a rhombus.

Method 2 Show that the quadrilateral is a parallelogram and that the diagonals are perpendicular.

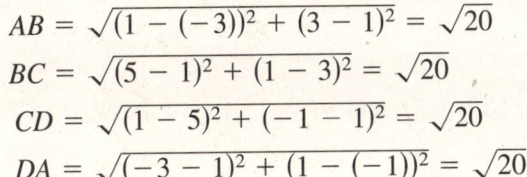

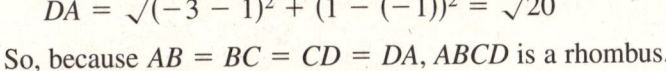

So, quadrilateral *ABCD* is a parallelogram because opposite sides are parallel.

LESSON 6.6 CONTINUED

Practice with Examples
For use with pages 364–370

$$\text{Slope of } \overline{AC} = \frac{1-1}{5-(-3)} = \frac{0}{8} = 0 \quad \text{(horizontal line)}$$

$$\text{Slope of } \overline{BD} = \frac{3-(-1)}{1-1} = \frac{4}{0} = \text{undefined} \quad \text{(vertical line)}$$

So, because the diagonals are perpendicular (in this case, one is horizontal and the other is vertical—more generally, the slope of one would be the negative reciprocal of the other's slope) and $ABCD$ is a parallelogram, $ABCD$ is a rhombus.

Exercises for Example 1

In Exercises 1 and 2, show that the quadrilateral with given vertices is a rhombus by using the methods demonstrated in Example 1.

1. $A(-4, -1)$, $B(-4, 2)$, $C(-1, 2)$, and $D(-1, -1)$. Use Method 1 from Example 1.

2. $E(-1, -2)$, $F(3, -1)$, $G(7, -2)$, and $H(3, -3)$. Use Method 2 from Example 1.

LESSON 6.6 CONTINUED

Practice with Examples
For use with pages 364–370

EXAMPLE 2 Identifying Special Quadrilaterals

For $P(0, 4)$, $Q(-4, 5)$, $R(-5, -1)$, and $S(-1, -2)$, what kind of quadrilateral is $PQRS$?

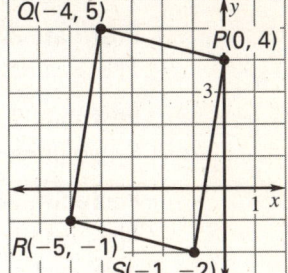

SOLUTION

Plot P, Q, R, and S in a coordinate plane.

Draw segments $\overline{PQ}$, $\overline{QR}$, $\overline{RS}$, and $\overline{SP}$.

$PQRS$ looks like a rectangle. Begin by seeing if $PQRS$ is a parallelogram.

Slope of $\overline{PQ} = \dfrac{5-4}{-4-0} = \dfrac{1}{-4} = -\dfrac{1}{4}$ Slope of $\overline{RS} = \dfrac{-2-(-1)}{-1-(-5)} = \dfrac{-1}{4} = -\dfrac{1}{4}$

Slope of $\overline{QR} = \dfrac{-1-5}{-5-(-4)} = \dfrac{-6}{-1} = 6$ Slope of $\overline{SP} = \dfrac{4-(-2)}{0-(-1)} = \dfrac{6}{1} = 6$

So, because the slopes are equal, $\overline{PQ} \parallel \overline{RS}$ and $\overline{QR} \parallel \overline{SP}$. Therefore $PQRS$ is a parallelogram. Next, see if the diagonals are congruent.

$QS = \sqrt{(-1-(-4))^2 + (-2-5)^2} = \sqrt{58}$
$PR = \sqrt{(-5-0)^2 + (-1-4)^2} = \sqrt{50}$

The diagonals are not congruent, so $PQRS$ is not a rectangle.

Thus, our conclusion about $PQRS$ is that it is a parallelogram.

Exercises for Example 2

Given coordinates for P, Q, R, and S, what kind of quadrilateral is $PQRS$?

3. $P(-2, 1)$, $Q(-2, 3)$, $R(3, 6)$, $S(0, 1)$

4. $P(0, 0)$, $Q(4, 0)$, $R(3, 7)$, $S(1, 7)$

5. $P(-1, -3)$, $Q(4, -3)$, $R(4, 3)$, $S(-1, 3)$

LESSON 6.7

Practice with Examples

For use with pages 372–380

GOAL Find the areas of squares, rectangles, parallelograms, and triangles and find the areas of trapezoids, kites, and rhombuses

---

Postulate 22 Area of a Square Postulate
The area of a square is the square of the length of its side.

Postulate 23 Area Congruence Postulate
If two polygons are congruent, then they have the same area.

Postulate 24 Area Addition Postulate
The area of a region is the sum of the areas of its nonoverlapping parts.

Theorem 6.20 Area of a Rectangle
The area of a rectangle is the product of its base and height.

Theorem 6.21 Area of a Parallelogram
The area of a parallelogram is the product of a base and its corresponding height.

Theorem 6.22 Area of a Triangle
The area of a triangle is one half the product of a base and its corresponding height.

Theorem 6.23 Area of a Trapezoid
The area of a trapezoid is one half the product of the height and the sum of the bases.

Theorem 6.24 Area of a Kite
The area of a kite is one half the product of the lengths of its diagonals.

Theorem 6.25 Area of a Rhombus
The area of a rhombus is equal to one half the product of the lengths of the diagonals.

EXAMPLE 1 Using the Area Theorems

a. Find the area of △ABC.

b. Find the area of ▱DEFG.

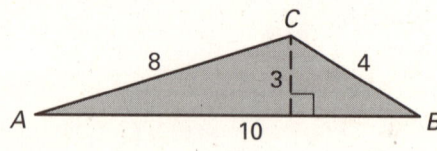
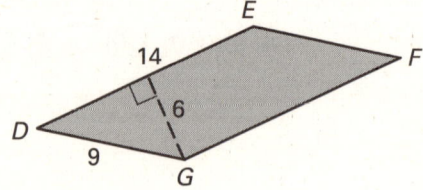

c. Find the area of trapezoid WXYZ.

d. Find the area of kite ABCD.
DB = 6, AC = 8

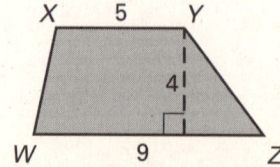
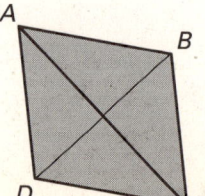

LESSON 6.7 CONTINUED

Practice with Examples
For use with pages 372–380

SOLUTION

a. Any one of the three sides of the triangle can be used as the base, b. Since the height of 3, which corresponds to the base $\overline{AB}$, is given, use $\overline{AB}$ for b.

Area $= \frac{1}{2}bh = \frac{1}{2}(10)(3) = 15$ square units

b. We use $\overline{DE}$ as the base because the corresponding height of 6 is given.
Area $= bh = (14)(6) = 84$ square units

c. Area $= \frac{1}{2}h(b_1 + b_2) = \frac{1}{2}(4)(5 + 9) = 28$ square units

d. Area $= \frac{1}{2}d_1 d_2 = \frac{1}{2}(6)(8) = 24$ square units

Exercises for Example 1

Find the area of the polygon.

1.

2.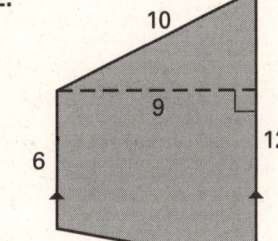

LESSON 6.7 CONTINUED

Practice with Examples
For use with pages 372–380

3.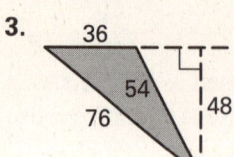

4.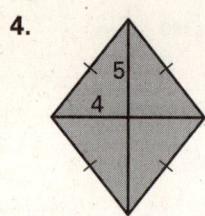

5.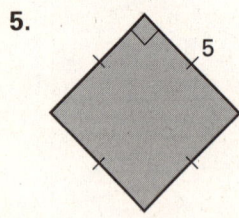

6.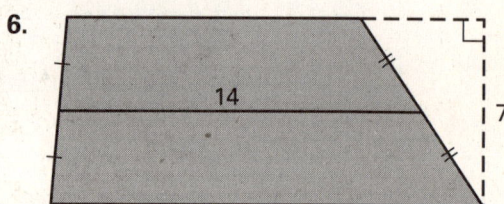

LESSON 7.1

Practice with Examples
For use with pages 396–402

GOAL Identify the three basic rigid transformations.

VOCABULARY

Figures in a plane can be reflected, rotated, or translated to produce new figures. The new figure is called the **image,** and the original figure is called the **preimage.**

The operation that *maps*, or moves, the preimage onto the image is called a **transformation.**

An **isometry** is a transformation that preserves lengths.

EXAMPLE 1 Naming Transformations

Use the graph of the transformation at the right.

a. Name and describe the transformation.

b. Name the coordinates of the vertices of the image.

c. Is △ABC congruent to its image?

SOLUTION

a. The transformation is a reflection in the y-axis.

b. The coordinates of the vertices of the image, △A′B′C, are A′(−4, 0), B′(−4, 4), and C(0, 4).

c. Yes, △ABC is congruent to its image △A′B′C. One way to show this would be to use the Distance Formula to find the lengths of the sides of both triangles. Then use the SSS Congruence Postulate.

Exercises for Example 1

In Exercises 1–3, use the graph of the transformation to answer the questions.

1. Name and describe the transformation.

2. Name the coordinates of the vertices of the image.

3. Name two angles with the same measure.

LESSON 7.1 CONTINUED

NAME _____ DATE _____

Practice with Examples

For use with pages 396–402

EXAMPLE 2 Identifying Isometries

Which of the following transformations appear to be isometries?

a. 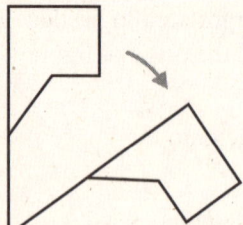   b. 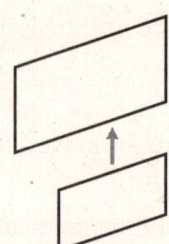   c.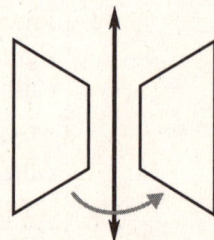

SOLUTION

a. This transformation appears to be an isometry. The preimage on the left is rotated about a point to produce a congruent image on the right.

b. This transformation is not an isometry. The parallelogram on the top is not congruent to its preimage on the bottom.

c. This transformation appears to be an isometry. The trapezoid on the left is reflected in a line to produce a congruent trapezoid on the right.

Exercises for Example 2

State whether the transformation appears to be an isometry.

4. 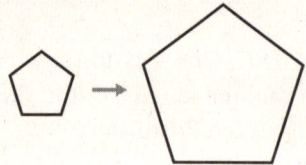   5.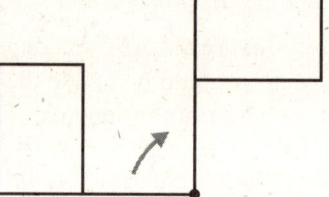

LESSON 7.1 CONTINUED

Practice with Examples
For use with pages 396–402

EXAMPLE 3 Preserving Length and Angle Measure

In the diagram, ABCD is mapped onto WXYZ. The mapping is a reflection. Given that $ABCD \rightarrow WXYZ$ is an isometry, find the length of $\overline{WX}$ and the measure of $\angle Y$.

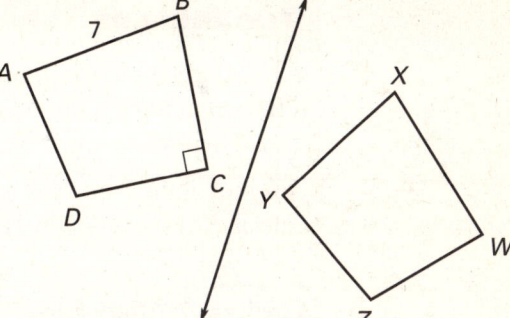

SOLUTION

Because the transformation is an isometry, the two figures are congruent. So, $WX = AB = 7$ and $m\angle Y = m\angle C = 90°$.

Exercises for Example 3

In Exercises 6 and 7, find the value of each variable, given that the transformation is an isometry.

6.

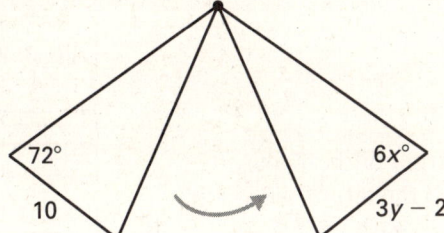

7.

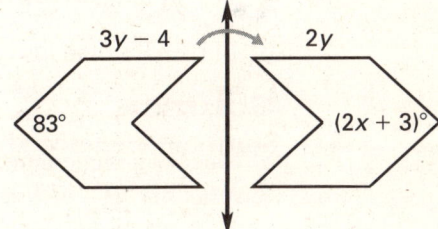

LESSON 7.2

Practice with Examples
For use with pages 404–410

GOAL Identify and use reflections in a plane and identify relationships between reflections and line symmetry.

VOCABULARY

A transformation which uses a line that acts like a mirror, with an image reflected in the line, is called a **reflection.** The line which acts like a mirror in a reflection is called the **line of reflection.**

A figure in the plane has a **line of symmetry** if the figure can be mapped onto itself by a reflection in the line.

Theorem 7.1 Reflection Theorem
A reflection is an isometry.

EXAMPLE 1 Reflections in a Coordinate Plane

Graph the given reflection.

a. $A(3, 2)$ in the y-axis

b. $B(1, -3)$ in the line $y = 1$

SOLUTION

a. Since A is three units to the right of the y-axis, its reflection, A', is three units to the left of the x-axis.

b. Start by graphing $y = 1$ and B. From the graph, you can see that B is 4 units below the line of reflection. This implies that its reflection, B', is 4 units above the line.

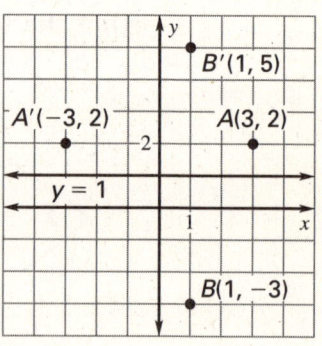

Exercises for Example 1

In Exercises 1–8, graph the given reflection.

1. $C(-1, 4)$ in the x-axis

2. $D(0, 3)$ in the y-axis

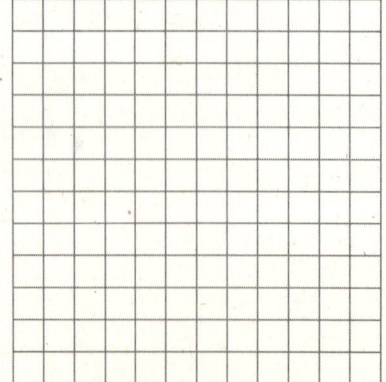
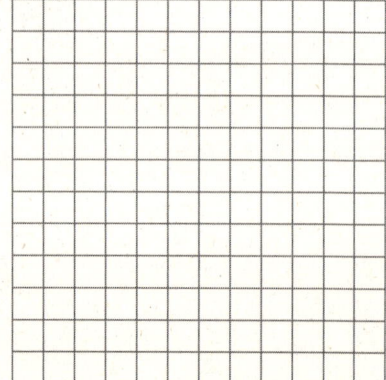

LESSON 7.2 CONTINUED

Practice with Examples
For use with pages 404–410

3. $E(4, -2)$ in the line $y = 3$

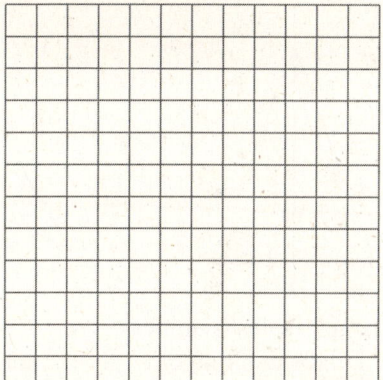

4. $F(1, -2)$ in the line $y = -2$

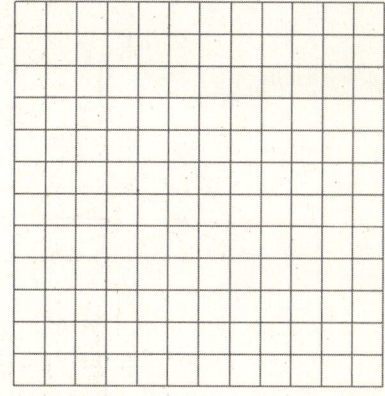

5. $G(3, 5)$ in the line $x = 1$

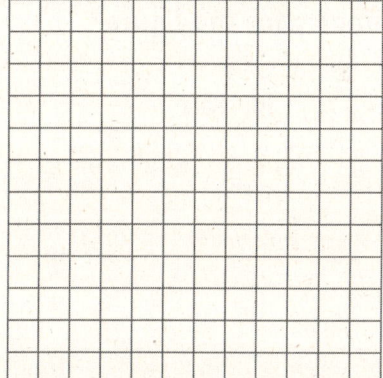

6. $H(-3, -1)$ in the line $x = 4$

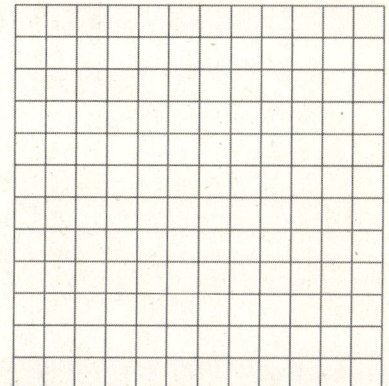

7. $I(4, 5)$ in the line $x = -2$

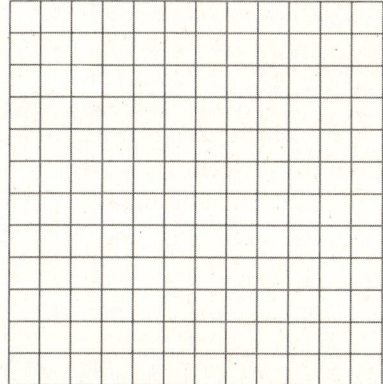

8. $J(-2, 3)$ in the line $y = 1$

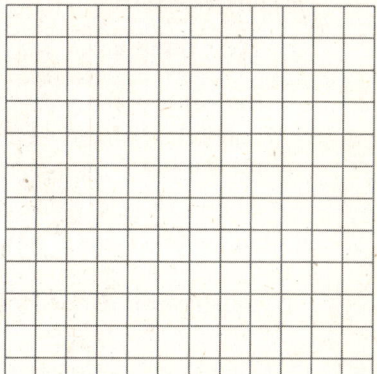

EXAMPLE 2 Finding Lines of Symmetry

Triangles can have different lines of symmetry depending on their shape. Find the number of lines of symmetry a triangle has when it is one of the following.

a. equilateral   **b.** isosceles   **c.** scalene

LESSON 7.2 CONTINUED

Practice with Examples
For use with pages 404–410

SOLUTION

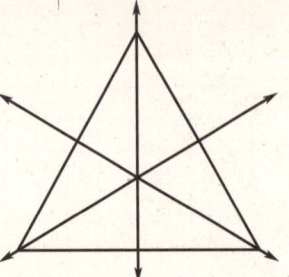
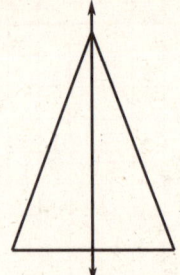
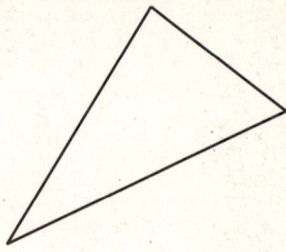

a. Equilateral triangles have three lines of symmetry.

b. Isosceles triangles have one line of symmetry.

c. Scalene triangles do not have any lines of symmetry.

Exercises for Example 2

Find the number of lines of symmetry for the figure described.

9. Rectangle

10. Kite

EXAMPLE 3 Finding a Minimum Distance

Find point C on the x-axis so $AC + BC$ is a minimum where A is $(-1, 5)$ and B is $(5, 1)$.

SOLUTION

Reflect A in the x-axis to obtain $A'(-1, -5)$. Then, draw $\overline{A'B}$. Label the point at which this segment intersects the x-axis as C. Because $\overline{A'B}$ represents the shortest distance between A' and B, and $AC = A'C$, you can conclude that at point C a minimum length is obtained. Next, to find the coordinates of C, find an equation for $\overline{A'B}$. Slope of $\overline{A'B} = \dfrac{1 - (-5)}{5 - (-1)} = \dfrac{6}{6} = 1$

Then use this slope and $A'(-1, -5)$ in $y - y_o = m(x - x_o)$ to get $y + 5 = x + 1$ or $y = x - 4$. Because C is on the x-axis, $y = 0$, so $x = 4$. Therefore, C is $(4, 0)$.

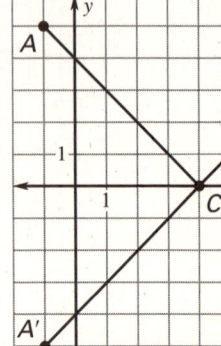

Exercises for Example 3

In Exercises 11–13, find point C on the x-axis so $AC + CB$ is a minimum.

11. $A(-1, -2)$, $B(8, -4)$

12. $A(1, 4)$, $B(8, 3)$

13. $A(-1.5, 6)$, $B(6, 9)$

LESSON 7.3

Practice with Examples
For use with pages 412–420

GOAL Identify rotations in a plane.

VOCABULARY

A **rotation** is a transformation in which a figure is turned about a fixed point.

The fixed point of a rotation is called the **center of rotation.**

Rays drawn from the center of rotation to a point and its image form an angle called the **angle of rotation.**

A figure in the plane has **rotational symmetry** if the figure can be mapped onto itself by a clockwise rotation of 180° or less.

Theorem 7.2 Rotation Theorem
A rotation is an isometry.

Theorem 7.3
If lines k and m intersect at point P, then a reflection in k followed by a reflection in m is a rotation about point P.

The angle of rotation is $2x°$, where $x°$ is the measure of the acute or right angle formed by k and m.

EXAMPLE 1 Rotations in a Coordinate Plane

In a coordinate plane, sketch the quadrilateral whose vertices are $A(-2, -1)$, $B(-5, 1)$, $C(-4, 5)$, and $D(-1, 2)$. Then, rotate $ABCD$ 90° clockwise about the origin and name the coordinates of the new vertices. Describe any patterns you see in the coordinates.

SOLUTION

Plot the points. Use a protractor, a compass, and a straightedge to find the rotated vertices. The coordinates of the preimage and image are listed below.

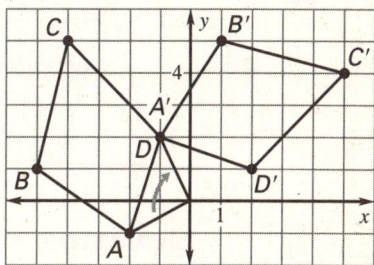

Figure $ABCD$	Figure $A'B'C'D'$
$A(-2, -1)$	$A'(-1, 2)$
$B(-5, 1)$	$B'(1, 5)$
$C(-4, 5)$	$C'(5, 4)$
$D(-1, 2)$	$D'(2, 1)$

In the list above, the x-coordinate of the image is the y-coordinate of the preimage. The y-coordinate of the image is the opposite of the x-coordinate of the preimage.

This transformation can be described as $(x, y) \rightarrow (y, -x)$.

LESSON 7.3 CONTINUED

NAME _____ DATE _____

Practice with Examples
For use with pages 412–420

Exercises for Example 1

In Exercises 1 and 2, use the given information to rotate the quadrilateral. Name the vertices of the image and compare with the vertices of the preimage. Describe any patterns you see.

1. 90° clockwise about origin

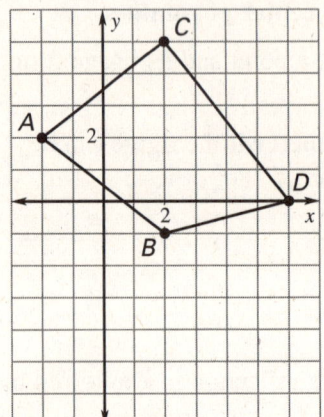

2. 180° counterclockwise about origin

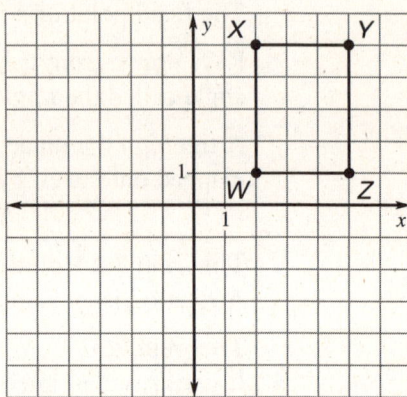

EXAMPLE 2 Identifying Rotational Symmetry

Which figures have rotational symmetry? For those that do, describe the rotations that map the figure onto itself.

a. Isosceles triangle   **b.** Kite   **c.** Rhombus

 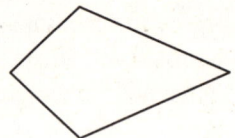 

SOLUTION

a. The isosceles triangle does not have rotational symmetry.

b. This kite has rotational symmetry. It can be mapped onto itself by a rotation of 180° about its center.

c. This rhombus has rotational symmetry. It can be mapped onto itself by a rotation of 180° about its center.

LESSON 7.3 CONTINUED

Practice with Examples
For use with pages 412–420

Exercises for Example 2

Decide which figures have rotational symmetry. For those that do, describe the rotations that map the figure onto itself.

3. Equilateral triangle

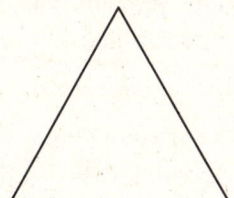

4. Rectangle

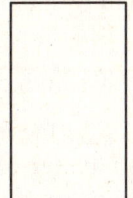

5. Regular pentagon

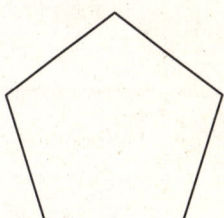

LESSON 7.4

Practice with Examples
For use with pages 421–428

GOAL Identify and use translations in the plane.

VOCABULARY

A **translation** is a transformation that maps every two points P and Q in the plane to points P' and Q', so that the following properties are true:
1) $PP' = QQ'$ and 2) $\overline{PP'} \parallel \overline{QQ'}$, or $\overline{PP'}$ and $\overline{QQ'}$ are collinear.

A **vector** is a quantity that has both direction and *magnitude*, or size.

When a vector is drawn as $\overrightarrow{PQ}$, the **initial point,** or starting point, of the vector is point P and the **terminal point,** or ending point, of the vector is point Q. $\overrightarrow{PQ}$ is read "vector PQ."

The **component form** of a vector combines the horizontal and vertical components.

Theorem 7.4 Translation Theorem
A translation is an isometry.

Theorem 7.5
If lines k and m are parallel, then a reflection in line k followed by a reflection in line m is a translation. If P'' is the image of P, then the following is true:

1. $\overleftrightarrow{PP''}$ is perpendicular to k and m.
2. $PP'' = 2d$, where d is the distance between k and m.

EXAMPLE 1 *Using Theorem 7.5*

In the diagram, a reflection in line k maps $\overline{AB}$ to $\overline{A'B'}$, a reflection in line m maps $\overline{A'B'}$ to $\overline{A''B''}$, $k \parallel m$, $AW = 7$, and $ZA'' = 3$.

a. Name some congruent segments.
b. Does $WZ = XY$? Explain.
c. What is the length of $\overline{BB''}$?

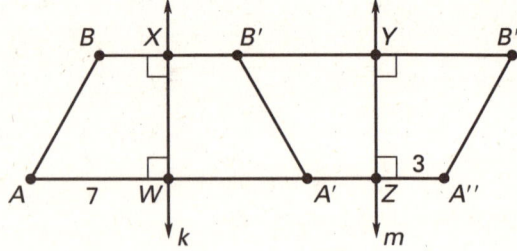

SOLUTION

a. Here are some sets of congruent segments: $\overline{AB}$, $\overline{A'B'}$, and $\overline{A''B''}$; $\overline{BX}$ and $\overline{XB'}$; $\overline{B'Y}$ and $\overline{YB''}$.

b. Yes, $WZ = XY$ because $\overline{WZ}$ and $\overline{XY}$ are opposite sides of a rectangle.

c. Because $BB'' = AA''$, the length of $\overline{BB''}$ is $7 + 7 + 3 + 3$, or 20 units.

LESSON 7.4 CONTINUED

Practice with Examples
For use with pages 421–428

Exercises for Example 1

In the diagram $k \parallel m$, $\triangle XYZ$ is reflected in line k, and $\triangle X'Y'Z'$ is reflected in line m.

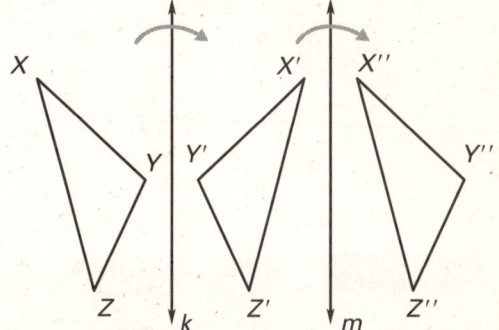

1. Name two segments parallel to $\overline{YY''}$.

2. If the length of $\overline{ZZ''}$ is 6 cm, what is the distance between k and m?

3. A translation maps $\triangle XYZ$ onto which triangle?

4. Which lines are perpendicular to $\overline{XX''}$?

EXAMPLE 2 — Translations in a Coordinate Plane

Sketch a quadrilateral with vertices $A(0, 4)$, $B(-2, 1)$, $C(0, -3)$, and $D(3, 4)$. Then sketch the image of the quadrilateral after the translation $(x, y) \rightarrow (x + 2, y - 1)$.

SOLUTION

Plot the points as shown. Shift each point 2 units to the right and 1 unit down to find the translated vertices.

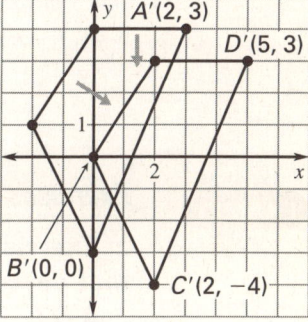

Exercises for Example 2

In Exercises 5–8, copy figure *PQRS* and draw its image after the translation.

5. $(x, y) \rightarrow (x - 4, y + 1)$ 6. $(x, y) \rightarrow (x, y - 5)$

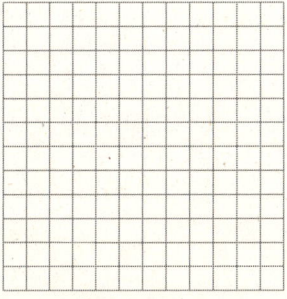 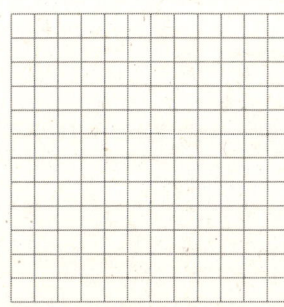

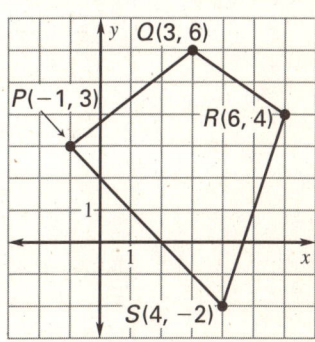

LESSON 7.4 CONTINUED

Practice with Examples
For use with pages 421–428

7. $(x, y) \rightarrow (x - 2, y - 2)$

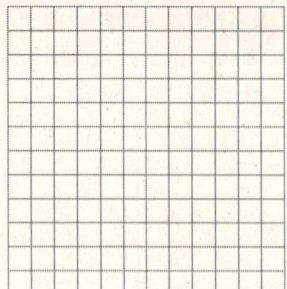

8. $(x, y) \rightarrow (x + 7, y + 3)$

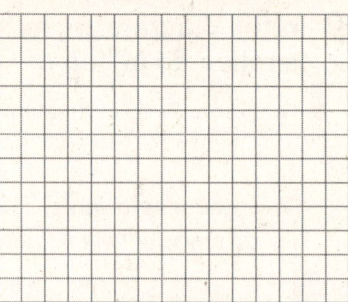

EXAMPLE 3 Finding Vectors

In the diagram, $\triangle ABC$ maps onto $\triangle A'B'C'$ by a translation. Write the component form of the vector that can be used to describe the translation.

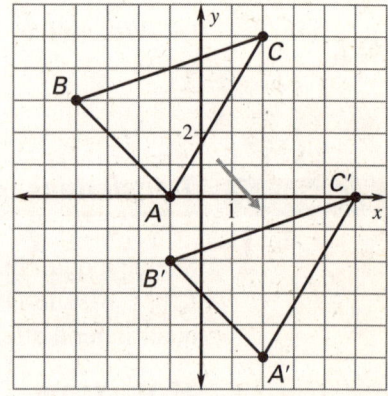

SOLUTION

Choose any vertex and its image, say A and A'. To move from A to A', you move 3 units to the right and 5 units down. The component form of the vector is $\langle 3, -5 \rangle$.

Exercises for Example 3

In Exercises 9 and 10, write the component form of the vector that describes the translation which maps $\triangle ABC$ onto $\triangle A'B'C'$.

9. $A(3, 6), B(1, 0), C(4, 8); A'(1, 2), B'(-1, -4), C'(2, 4)$

10. $A(-6, -2), B(-5, 3), C(1, -1); A'(-3, -5), B'(-2, 0), C'(4, -4)$

132 Geometry
Practice Workbook with Examples

LESSON 7.5

NAME _____ DATE _____

Practice with Examples
For use with pages 430–436

GOAL Identify glide reflections in a plane and represent transformations as compositions of simpler transformations.

VOCABULARY

A **glide reflection** is a transformation in which every point P is mapped onto a point P'' by the following steps:

1. A translation maps P onto P'.

2. A reflection in a line k parallel to the direction of the translation maps P' onto P''.

When two or more transformations are combined to produce a single transformation, the result is called a **composition** of the transformations.

Theorem 7.6 Composition Theorem
The composition of two (or more) isometries is an isometry.

EXAMPLE 1 Finding the Image of a Glide Reflection

Use the information below to sketch the image of $\triangle ABC$ after a glide reflection.

$A(-5, 5)$, $B(-3, 2)$, $C(1, 5)$

Translation: $(x, y) \rightarrow (x, y - 7)$

Reflection: in the line $x = 3$

SOLUTION

Begin by graphing $\triangle ABC$. Then, shift the triangle 7 units down to produce $\triangle A'B'C'$. Finally, reflect the triangle in the line $x = 3$ to produce $\triangle A''B''C''$.

Coordinates		
$\triangle ABC$	$\triangle A'B'C'$	$\triangle A''B''C''$
$A(-5, 5)$	$A'(-5, -2)$	$A''(11, -2)$
$B(-3, 2)$	$B'(-3, -5)$	$B''(9, -5)$
$C(1, 5)$	$C'(1, -2)$	$C''(5, -2)$

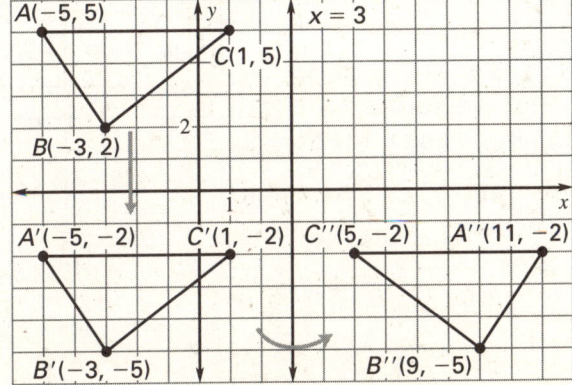

LESSON 7.5 CONTINUED

Practice with Examples
For use with pages 430–436

Exercises for Example 1

Sketch the image of △ABC after a glide reflection using the given transformations in the order they appear. Then, reverse the order of the transformations and sketch the image again. Determine if the order of the transformations affects the image.

1. $A(0, 0)$, $B(0, 5)$, $C(7, 0)$
   Translation: $(x, y) \rightarrow (x + 3, y)$
   Reflection: in the x-axis

2. $A(-3, 2)$, $B(-1, -2)$, $C(3, 2)$
   Translation: $(x, y) \rightarrow (x - 4, y + 2)$
   Reflection: in the line $x = 2$

3. $A(3, -1)$, $B(7, -1)$, $C(6, 2)$
   Translation: $(x, y) \rightarrow (x - 1, y + 5)$
   Reflection: in the line $y = -1$

4. $A(-4, 0)$, $B(0, 7)$, $C(3, 1)$
   Translation: $(x, y) \rightarrow (x, y + 3)$
   Reflection: in the line $x = 4$

EXAMPLE 2 Finding the Image of a Composition

Sketch the image of $\overline{AB}$ after a composition of the given rotation and reflection.

$A(-5, 3)$, $B(-3, 7)$

Rotation: 90° clockwise about the origin

Reflection: in the line $y = 1$

SOLUTION

Begin by graphing $\overline{AB}$. Then rotate the segment 90° clockwise about the origin to produce $\overline{A'B'}$. Finally, reflect the segment in the line $y = 1$ to produce $\overline{A''B''}$.

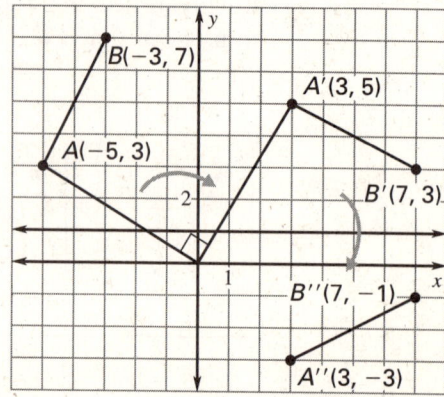

LESSON 7.5 CONTINUED

NAME _____ DATE _____

Practice with Examples
For use with pages 430–436

Exercises for Example 2

Sketch the image of $\overline{AB}$ after a composition using the given transformations.

5. $A(-5, 5), B(-3, 2)$

 Translation: $(x, y) \rightarrow (x + 8, y - 2)$
 Reflection: in the x-axis

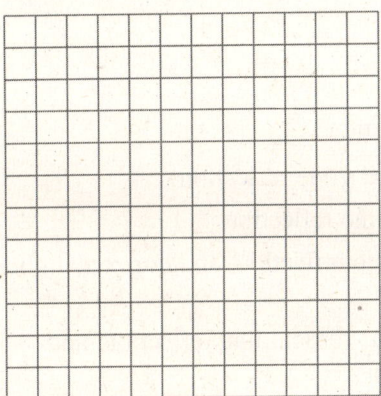

6. $A(0, -8), B(3, -4)$

 Rotation: 180° clockwise about the origin
 Reflection: in the line $x = 3$

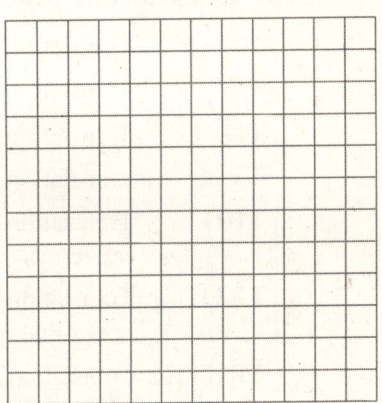

7. $A(6, -1), B(9, 4)$

 Translation: $(x, y) \rightarrow (x - 8, y + 1)$
 Reflection: in the y-axis

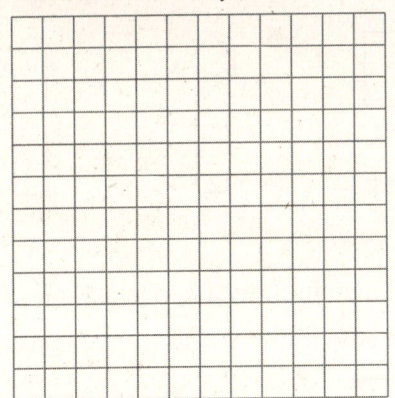

8. $A(3, 10), B(7, 5)$

 Translation: $(x, y) \rightarrow (x - 4, y)$
 Rotation: 90° counterclockwise about the origin

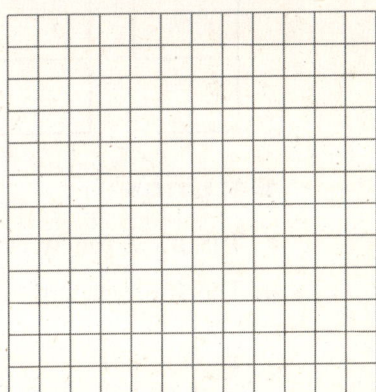

LESSON 7.6

Practice with Examples
For use with pages 437–443

GOAL Use transformations to classify frieze patterns.

VOCABULARY

A **frieze pattern** or **border pattern** is a pattern that extends to the left and right in such a way that the pattern can be mapped onto itself by a horizontal translation.

Classification of Frieze Patterns

T	Translation
TR	Translation and 180° rotation
TG	Translation and horizontal glide reflection
TV	Translation and vertical line reflection
THG	Translation, horizontal line reflection, and horizontal glide reflection
TRVG	Translation, 180° rotation, vertical line reflection, and horizontal glide reflection
TRHVG	Translation, 180° rotation, horizontal line reflection, vertical line reflection, and horizontal glide reflection

EXAMPLE 1 Classifying Patterns

Name the isometries that map the frieze pattern onto itself.

a. 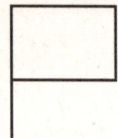 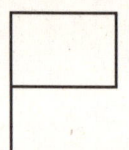 b.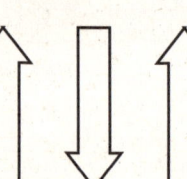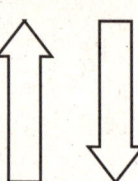

SOLUTION

a. This frieze pattern can be mapped onto itself by a horizontal translation (T).

b. This frieze pattern can be mapped onto itself by a horizontal translation (T)

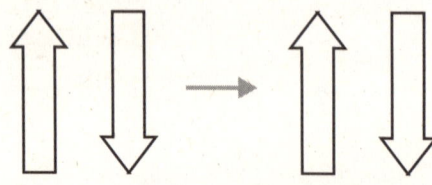

or by a horizontal glide reflection (G).

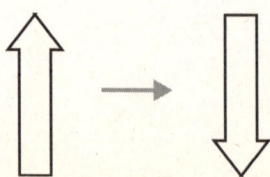

LESSON 7.6 CONTINUED

NAME _____ DATE _____

Practice with Examples
For use with pages 437–443

Exercises for Example 1

In Exercises 1–5, name the isometries that map the frieze pattern onto itself.

1. 2.

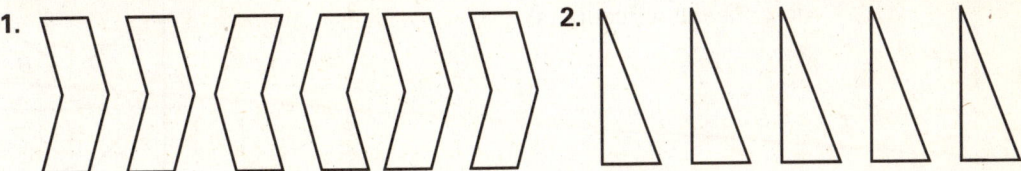

3.

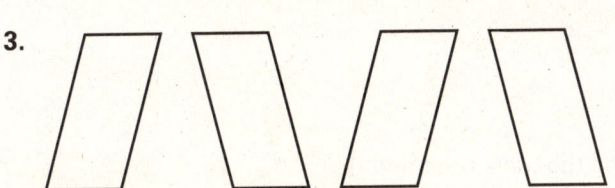

EXAMPLE 2 Describing Transformations

Use the diagram of the frieze pattern.

a. Is there a reflection in a vertical line?

b. Is there a reflection in a horizontal line?

c. Name and describe the transformation that maps A onto F.

d. Name and describe the transformation that maps D onto E.

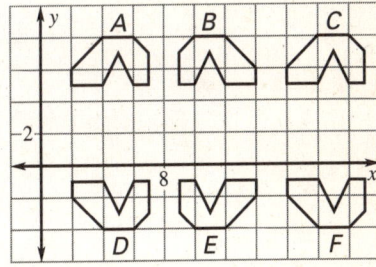

SOLUTION

a. Yes, there is a reflection in the line $x = 8$ and also in the line $x = 15$.

b. Yes, there is a reflection in the line $y = 2$.

c. A can be mapped onto F by a horizontal glide reflection.

d. D can be mapped onto E by a translation.

LESSON 7.6 CONTINUED

NAME _____ DATE _____

Practice with Examples

For use with pages 437–443

Exercises for Example 2

In Exercises 4–7, use the diagram of the frieze pattern.

4. Is there a reflection in a horizontal line? If so, describe the reflection(s).

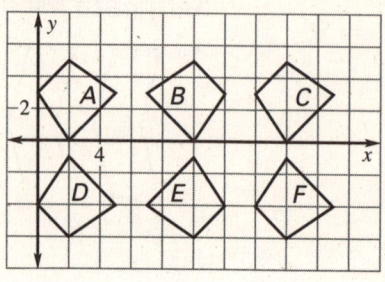

5. Is there a reflection in a vertical line? If so, describe the reflection(s).

6. Name and describe the transformation that maps B onto C.

7. Name and describe the transformation that maps D onto C.

LESSON 8.1

Practice with Examples
For use with pages 457–464

GOAL Find and simplify the ratio of two numbers

VOCABULARY

If a and b are two quantities that are measured in the same units, then the **ratio of a to b** is $\dfrac{a}{b}$.

An equation that equates two ratios is a **proportion**.

In the proportion $\dfrac{a}{b} = \dfrac{c}{d}$, the numbers a and d are the **extremes** of the proportion and the numbers b and c are the **means** of the proportion.

Properties of Proportions

1. **Cross Product Property**   The product of the extremes equals the product of the means.

    If $\dfrac{a}{b} = \dfrac{c}{d}$, then $ad = bc$.

2. **Reciprocal Property**   If two ratios are equal, then their reciprocals are also equal.

    If $\dfrac{a}{b} = \dfrac{c}{d}$, then $\dfrac{b}{a} = \dfrac{d}{c}$.

EXAMPLE 1 Simplifying Ratios

Simplify the ratios.

a. $\dfrac{8 \text{ in.}}{2 \text{ ft}}$

b. $\dfrac{1 \text{ km}}{500 \text{ m}}$

SOLUTION

To simplify ratios with unlike units, convert to like units so that the units divide out. Then simplify the fraction, if possible.

a. $\dfrac{8 \text{ in.}}{2 \text{ ft}} = \dfrac{8 \text{ in.}}{2 \cdot 12 \text{ in.}} = \dfrac{8}{24} = \dfrac{1}{3}$

b. $\dfrac{1 \text{ km}}{500 \text{ m}} = \dfrac{1 \cdot 1000 \text{ m}}{500 \text{ m}} = \dfrac{1000}{500} = \dfrac{2}{1}$

LESSON 8.1 CONTINUED

Practice with Examples

For use with pages 457–464

Exercises for Example 1

Simplify the ratio.

1. $\dfrac{25 \text{ cm}}{2 \text{ m}}$
2. $\dfrac{18 \text{ ft}}{2 \text{ yd}}$
3. $\dfrac{2 \text{ ft}}{24 \text{ in.}}$
4. $\dfrac{6 \text{ km}}{9 \text{ km}}$

EXAMPLE 2 Using Ratios

Triangle XYZ has an area of 25 square inches. The ratio of the base of $\triangle XYZ$ to the height of $\triangle XYZ$ is 2:1. Find the base and height of $\triangle XYZ$.

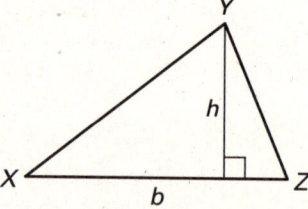

SOLUTION

Because the ratio of the base to the height is 2:1, you can represent the base as $2h$.

$A = \dfrac{1}{2}bh$ Formula for the area of a triangle

$25 = \dfrac{1}{2}(2h)h$ Substitute for A and b.

$25 = h^2$ Simplify.

$5 = h$ Find the positive square root.

So, $\triangle XYZ$ has a base of 10 inches and a height of 5 inches.

Exercise for Example 2

5. The area of a rectangle is 125 ft². The ratio of the width to the length is 1:5. Find the length and the width.

LESSON 8.1 CONTINUED

NAME _____ DATE _____

Practice with Examples
For use with pages 457–464

EXAMPLE 3 Solving Proportions

Solve the proportion.

$\dfrac{x}{8} = \dfrac{5}{4}$

SOLUTION

$\dfrac{x}{8} = \dfrac{5}{4}$	Write original proportion.
$4x = 40$	Cross product property
$x = 10$	Divide each side by 4.

Exercises for Example 3

Find the value of x by solving the proportion.

6. $\dfrac{9}{x} = \dfrac{2}{7}$ 7. $\dfrac{5}{3} = \dfrac{5x}{6}$ 8. $\dfrac{4}{x-4} = \dfrac{3}{x}$ 9. $\dfrac{3}{x} = \dfrac{x}{12}$

LESSON 8.2

Practice with Examples

For use with pages 465–471

GOAL Use properties of proportions

VOCABULARY

The **geometric mean** of two positive numbers a and b is the positive number x such that $\dfrac{a}{x} = \dfrac{x}{b}$.

Additional Properties of Proportions

3. If $\dfrac{a}{b} = \dfrac{c}{d}$, then $\dfrac{a}{c} = \dfrac{b}{d}$

4. If $\dfrac{a}{b} = \dfrac{c}{d}$, then $\dfrac{a+b}{b} = \dfrac{c+d}{d}$

EXAMPLE 1 Using Properties of Proportions

In the diagram $\dfrac{BC}{DC} = \dfrac{AC}{EC}$. Find the lengths of $\overline{DC}$ and $\overline{BC}$.

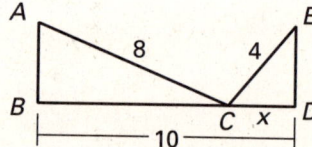

SOLUTION

Let $DC = x$. Then $BC = 10 - x$.

$\dfrac{BC}{DC} = \dfrac{AC}{EC}$ Given

$\dfrac{10 - x}{x} = \dfrac{8}{4}$ Substitute.

$\dfrac{10 - x}{x} = \dfrac{2}{1}$ Simplify.

$10 - x = 2x$ Cross product property

$10 = 3x$ Add x to each side.

$\dfrac{10}{3} = x$ Divide each side by 3.

$x \approx 3.3$ Use a calculator.

So, $DC \approx 3.3$ and $BD \approx 10 - 3.3 = 6.7$.

LESSON 8.2 CONTINUED

Practice with Examples
For use with pages 465–471

Exercises for Example 1
Find the value of x.

1. $\dfrac{AB}{BC} = \dfrac{FE}{ED}$

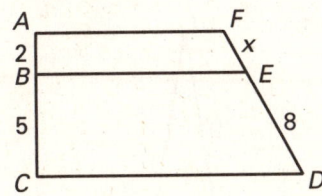

2. $\dfrac{BG}{CD} = \dfrac{AB}{AC}$

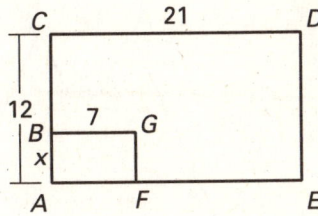

EXAMPLE 2 Using a Geometric Mean

In the diagram $\dfrac{AC}{DC} = \dfrac{DC}{BC}$. Find the value of x.

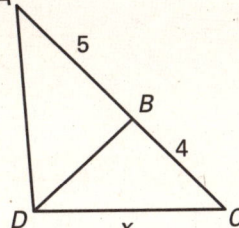

SOLUTION

Let $DC = x$.

$\dfrac{AC}{x} = \dfrac{x}{BC}$ Write proportion.

$\dfrac{9}{x} = \dfrac{x}{4}$ Substitute.

$x^2 = 9 \cdot 4$ Cross product property

$x = \sqrt{36} = 6$ Simplify.

LESSON 8.2 CONTINUED

Practice with Examples
For use with pages 465–471

Exercises for Example 2
Find the value of x.

3. $\dfrac{CD}{AB} = \dfrac{AB}{EF}$

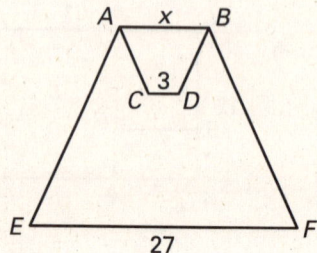

4. $\dfrac{AB}{AD} = \dfrac{AD}{DC}$

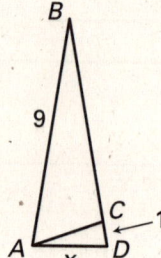

LESSON 8.3

NAME _____ DATE _____

Practice with Examples

For use with pages 473–479

GOAL Identify and use similar polygons

VOCABULARY

When there is a correspondence between two polygons such that their corresponding angles are congruent and the lengths of corresponding sides are proportional the two polygons are called **similar polygons.**

Theorem 8.1 If two polygons are similar, then the ratio of their perimeters is equal to the ratios of their corresponding side lengths.

EXAMPLE 1 *Writing Similarity Statements*

Quadrilaterals *ABCD* and *EFGH* are similar. List all the pairs of congruent angles. Write the ratios of the corresponding sides in a statement of proportionality.

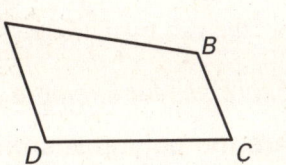

SOLUTION

Because $ABCD \sim EFGH$ you can write $\angle A \cong \angle E$, $\angle B \cong \angle F$, $\angle C \cong \angle G$, and $\angle D \cong \angle H$. You can write the statement of proportionality as follows:

$$\frac{AB}{EF} = \frac{BC}{FG} = \frac{CD}{GH} = \frac{DA}{HE}.$$

Exercises for Example 1

The two polygons are similar. List all the pairs of congruent angles. Write the ratios of the corresponding sides in a statement of proportionality.

1. △*ABC* ~ △*DEF*

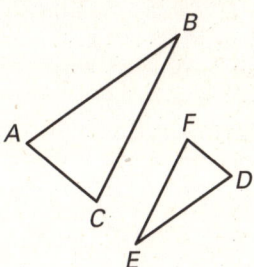

2. *ABDC* ~ *ZWXY*

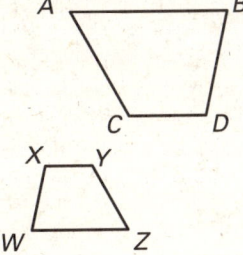

3. *EFGHJ* ~ *MRQPN*

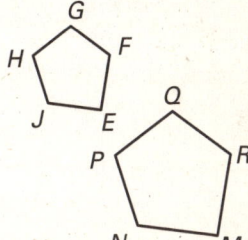

LESSON 8.3 CONTINUED

Practice with Examples
For use with pages 473–479

EXAMPLE 2 Comparing Similar Polygons

Decide whether the figures are similar. If they are similar, write a similarity statement.

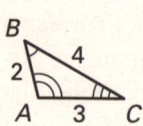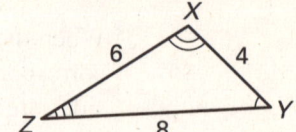

SOLUTION

The corresponding angles of △ABC and △XYZ are congruent. Also, the corresponding side lengths are proportional.

$$\frac{AB}{XY} = \frac{2}{4} = \frac{1}{2} \qquad \frac{BC}{YZ} = \frac{4}{8} = \frac{1}{2} \qquad \frac{CA}{ZX} = \frac{3}{6} = \frac{1}{2}$$

So, the two triangles are similar and you can write △ABC ~ △XYZ.

Exercises for Example 2

Are the polygons similar? If so, write a similarity statement.

4.
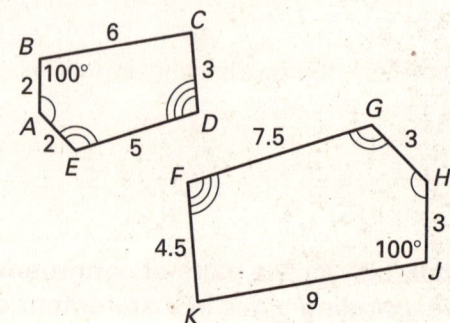

5.
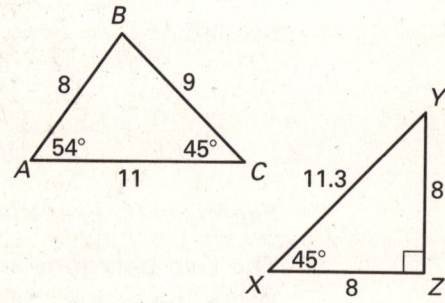

LESSON 8.3 CONTINUED

Practice with Examples
For use with pages 473–479

EXAMPLE 3 Using Similar Polygons

Pentagon *ABCDE* is similar to pentagon *JKLMN*. Find the value of x.

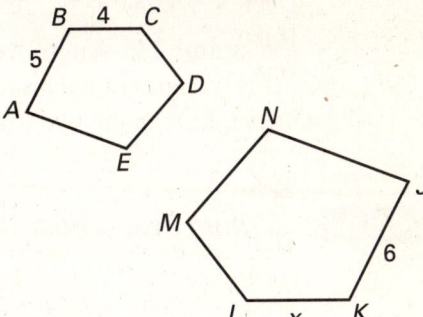

SOLUTION

Set up a proportion that contains *KL*.

$\dfrac{AB}{JK} = \dfrac{BC}{KL}$ Write a proportion.

$\dfrac{5}{6} = \dfrac{4}{x}$ Substitute.

$x = 4.8$ Cross multiply and divide by 5.

Exercises for Example 3

Find the value of x.

6. *ABCD* ~ *WXYZ*

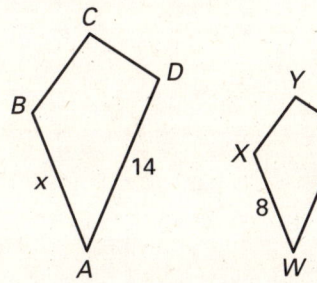

7. *JKLMN* ~ *PQRST*

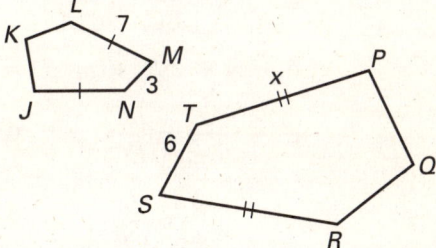

LESSON 8.4

Practice with Examples
For use with pages 480–487

GOAL Identify similar triangles

VOCABULARY

Postulate 25 Angle-Angle (AA) Similarity Postulate
If two angles of one triangle are congruent to two angles of another triangle, then the two triangles are similar.

EXAMPLE 1 Writing Proportionality Statements

In the diagram $\triangle ABC \sim \triangle DEC$.

  a. Write the statement of proportionality.
  b. Find $m\angle D$.
  c. Find the length of $\overline{CE}$.

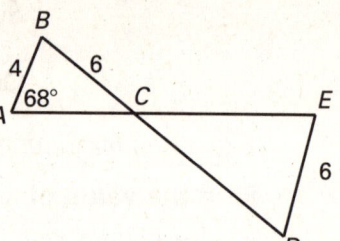

SOLUTION

a. $\dfrac{AB}{DE} = \dfrac{BC}{EC} = \dfrac{CA}{CD}$

b. $\angle A \cong \angle D$, so $m\angle D = 68°$.

c. $\dfrac{AB}{DE} = \dfrac{BC}{EC}$ Write proportion.

$\dfrac{4}{6} = \dfrac{6}{CE}$ Substitute.

$4 \cdot CE = 36$ Cross product property

$CE = 9$ Divide each side by 4.

So, the length of $\overline{CE}$ is 9.

LESSON 8.4 CONTINUED

NAME _____ DATE _____

Practice with Examples
For use with pages 480–487

Exercises for Example 1

The triangles shown are similar. List all the pairs of congruent angles and write the statement of proportionality. Find the value of *x*.

1.
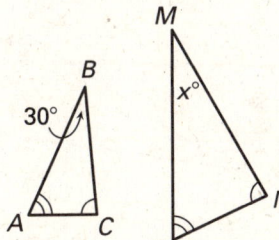

2.
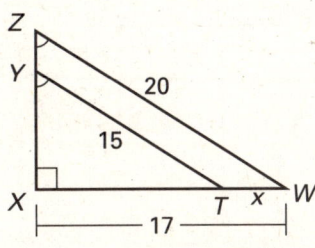

3.
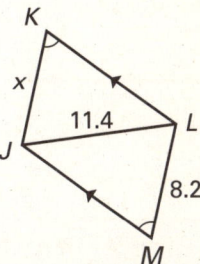

EXAMPLE 2 Proving that Two Triangles are Similar

Determine whether the triangles can be proved similar. If they are similar, write a similarity statement. If they are not similar, explain why.

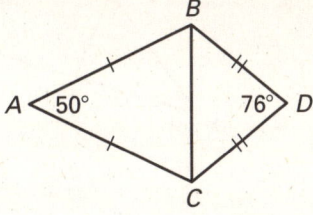

SOLUTION

In △ABC, you are given that $m\angle A = 50°$. Because △ABC is an isosceles triangle, you know that $m\angle ABC = m\angle ACB = \frac{180° - 50°}{2} = \frac{130°}{2} = 65°$. Similarly, you can find the angles of △DBC to be 76°, 52°, and 52°. Because the angles in △ABC are not congruent to the angles in △DBC, the triangles are not similar.

Geometry
Practice Workbook with Examples

LESSON 8.4 CONTINUED

Practice with Examples
For use with pages 480–487

Exercises for Example 2

Determine whether the triangles can be proved similar. If they are similar, write a similarity statement. If they are not similar, explain why.

4.
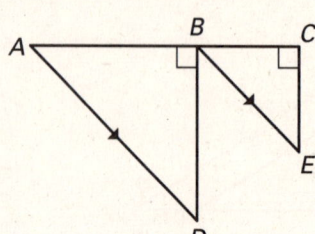

5.
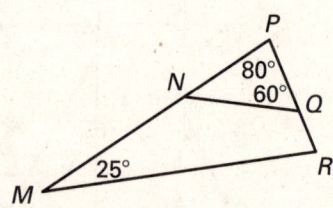

6.
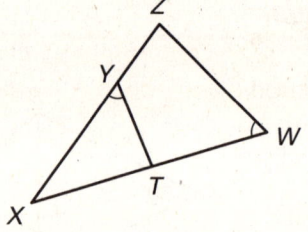

LESSON 8.5

NAME _____ DATE _____

Practice with Examples
For use with pages 488–496

GOAL Use similarity theorems to prove that two triangles are similar

> **VOCABULARY**
>
> **Theorem 8.2 Side-Side-Side (SSS) Similarity Theorem**
> If the lengths of the corresponding sides of two triangles are proportional, then the triangles are similar.
>
> **Theorem 8.3 Side-Angle-Side (SAS) Similarity Theorem**
> If an angle of one triangle is congruent to an angle of a second triangle and the lengths of the sides including these angles are proportional, then the triangles are similar.

EXAMPLE 1 Using the SSS Similarity Theorem

Which of the following triangles are similar?

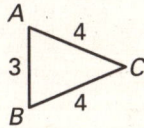 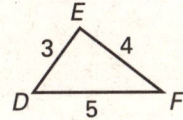 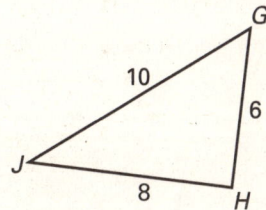

SOLUTION

To decide which, if any, of the triangles are similar, you need to consider the ratios of the lengths of corresponding sides.

Ratios of Side Lengths of △ABC and △DEF

$$\frac{AB}{DE} = \frac{3}{3} = \frac{1}{1} \qquad \frac{CA}{DF} = \frac{4}{5} \qquad \frac{BC}{EF} = \frac{4}{4} = \frac{1}{1}$$

Shortest sides   Longest sides   Remaining sides

Because the ratios are not equal, △ABC and △DEF are not similar.

Ratios of Side Lengths of △GHJ and △DEF

$$\frac{GH}{DE} = \frac{6}{3} = \frac{2}{1} \qquad \frac{GJ}{DF} = \frac{10}{5} = \frac{2}{1} \qquad \frac{HJ}{EF} = \frac{8}{4} = \frac{2}{1}$$

Shortest sides   Longest sides   Remaining sides

Because the ratios are equal, △GHJ ~ △DEF.

Since △DEF is similar to △GHJ and △DEF is not similar to △ABC, △GHJ is not similar to △ABC.

LESSON 8.5 CONTINUED

NAME _____ DATE _____

Reteaching with Practice

For use with pages 488–496

Exercises for Example 1

Determine which two of the three given triangles are similar.

1.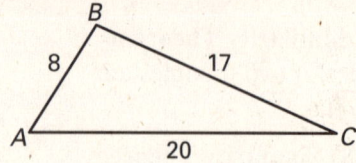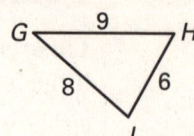

2.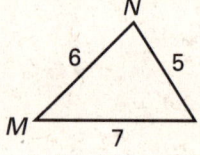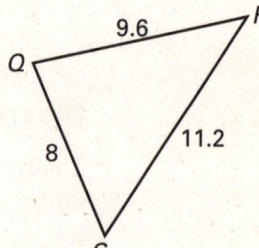

LESSON 8.5 CONTINUED

NAME _____ DATE _____

Practice with Examples
For use with pages 488–496

EXAMPLE 2 *Using the SAS Similarity Theorem*

Use the given lengths to prove that △ABC ~ △DEC.

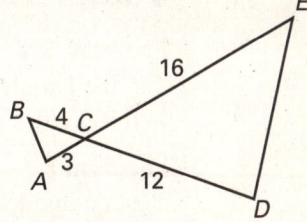

SOLUTION

Begin by finding the ratios of the lengths of the corresponding sides.

$$\frac{AC}{DC} = \frac{3}{12} = \frac{1}{4} \qquad \frac{BC}{EC} = \frac{4}{16} = \frac{1}{4}$$

So, the side lengths AC and BC of △ABC are proportional to the corresponding side lengths DC and EC of △DEC. The included angle in △ABC is ∠BCA; the included angle in △DEC is ∠ECD. Because these two angles are vertical angles, they are congruent. So, by the SAS Similarity Theorem, △ABC ~ △DEC.

Exercises for Example 2

Prove that the two triangles are similar.

3.

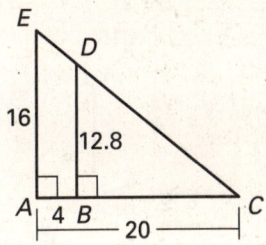

4.

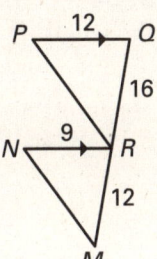

5.

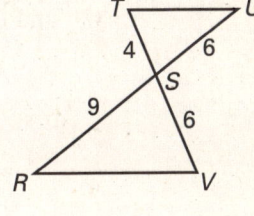

LESSON 8.6

Practice with Examples
For use with pages 498–505

GOAL Use proportionality theorems to calculate segment lengths

VOCABULARY

Theorem 8.4   Triangle Proportionality Theorem
If a line parallel to one side of a triangle intersects the other two sides, then it divides the two sides proportionally.

Theorem 8.5   Converse of the Triangle Proportionality Theorem
If a line divides two sides of a triangle proportionally, then it is parallel to the third side.

Theorem 8.6
If three parallel lines intersect two transversals, then they divide the transversals proportionally.

Theorem 8.7
If a ray bisects an angle of a triangle, then it divides the opposite side into segments whose lengths are proportional to the lengths of the other two sides.

EXAMPLE 1   Finding the Length of a Segment

a. In the diagram, $\overline{AB} \parallel \overline{CD}$, $BD = 15$, $AC = 10$, and $CE = 18$. What is the length of $\overline{DE}$?

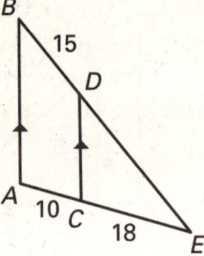

b. In the diagram, $\overline{NR} \parallel \overline{PQ}$, $MQ = 42$, $MN = 13$, and $NP = 8$. What is the length of $\overline{RQ}$ and $\overline{MR}$?

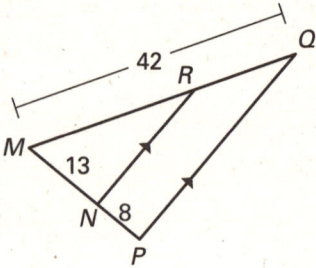

SOLUTION

a. $\dfrac{DE}{BD} = \dfrac{CE}{AC}$ Triangle Proportionality Theorem

$\dfrac{DE}{15} = \dfrac{18}{10}$ Substitute.

$DE = \dfrac{15(18)}{10} = 27$ Multiply each side by 15 and simplify.

154   Geometry
Practice Workbook with Examples

Practice with Examples

For use with pages 498–505

b. Let $RQ = x$. Then $MR = 42 - x$.

$\dfrac{MR}{RQ} = \dfrac{MN}{NP}$ Triangle Proportionality Theorem

$\dfrac{42 - x}{x} = \dfrac{13}{8}$ Substitute.

$8(42 - x) = 13x$ Cross product property

$336 - 8x = 13x$ Distributive property

$16 = x$ Simplify.

So, $RQ = 16$ and $MR = 42 - 16 = 26$.

Exercises for Example 1

Find the value of each variable.

1.
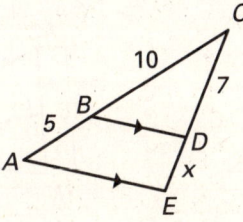

2.
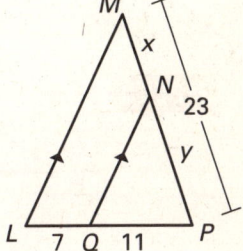

3.

LESSON 8.6 CONTINUED

NAME _____ DATE _____

Practice with Examples

For use with pages 498–505

EXAMPLE 2 *Using Proportionality Theorems*

In the diagram, $\overline{MP}$ bisects $\angle M$. Find NP.

SOLUTION

$\dfrac{NP}{PQ} = \dfrac{MN}{MQ}$ Apply Theorem 8.7.

$\dfrac{NP}{4} = \dfrac{16}{18}$ Substitute.

$NP = \dfrac{4(16)}{18} \approx 3.6$ Multiply each side by 4 and simplify.

Exercises for Example 2

Find the value of each variable.

4.
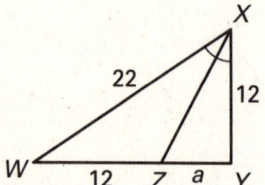

5.
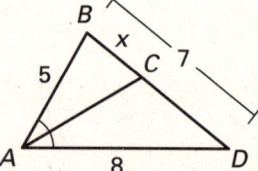

6.
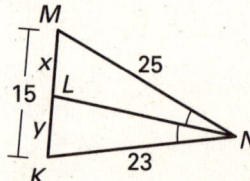

LESSON 8.7

Practice with Examples
For use with pages 506–513

GOAL Identify dilations and use properties of dilations to create a perspective drawing

VOCABULARY

A **dilation** with center C and scale factor k is a transformation that maps every point P in the plane to a point P' so that the following properties are true.

1. If P is not the center point C, then the image point P' lies on $\overrightarrow{CP}$. The scale factor k is a positive number such that $k = \dfrac{CP'}{CP}$, and $k \neq 1$.

2. If P is the center point C, then $P = P'$.

A dilation is a **reduction** if $0 < k < 1$.

A dilation is an **enlargement** if $k > 1$.

EXAMPLE 1 Identifying Dilations

Identify the dilation and find its scale factor.

a.
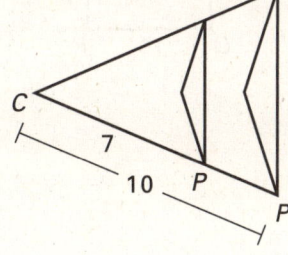

b.
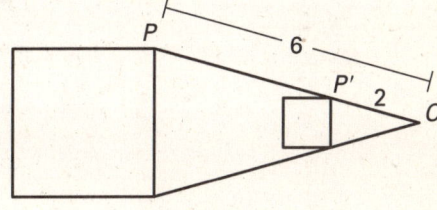

SOLUTION

a. Because $\dfrac{CP'}{CP} = \dfrac{10}{7}$, the scale factor is $k = \dfrac{10}{7}$. This is an enlargement.

b. Because $\dfrac{CP'}{CP} = \dfrac{2}{6} = \dfrac{1}{3}$, the scale factor is $k = \dfrac{1}{3}$. This is a reduction.

LESSON 8.7 CONTINUED

Practice with Examples

For use with pages 506–513

Exercises for Example 1

Identify the dilation and find its scale factor.

1.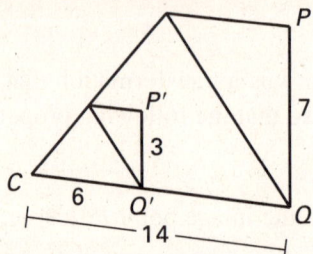

2.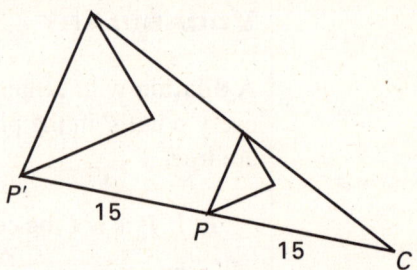

3.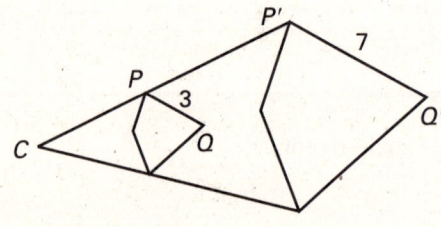

4.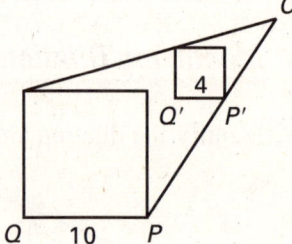

EXAMPLE 2 Dilation in a Coordinate Plane

Draw a dilation of $\triangle ABC$ with $A(1, 2)$, $B(5, 0)$, and $C(3, 4)$. Use the origin as the center and use a scale factor of $k = 2$.

SOLUTION

Because the origin is the center, you can find the image of each vertex by multiplying its coordinates by the scale factor.

$A(1, 2) \rightarrow A'(2, 4)$

$B(5, 0) \rightarrow B'(10, 0)$

$C(3, 4) \rightarrow C'(6, 8)$

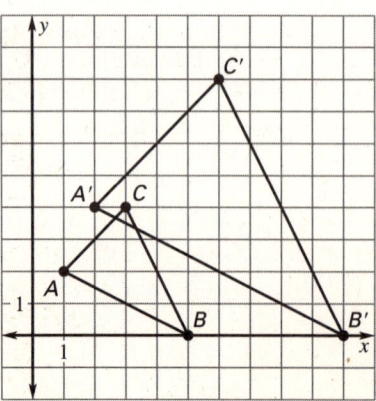

LESSON 8.7 CONTINUED

NAME _____ DATE _____

Practice with Examples
For use with pages 506–513

Exercises for Example 2

Use the origin as the center of the dilation and the given scale factor to find the coordinates of the vertices of the image of the polygon.

5. $k = \dfrac{3}{2}$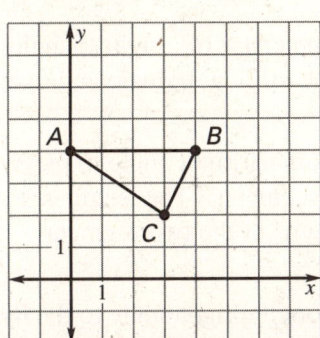

6. $k = 3$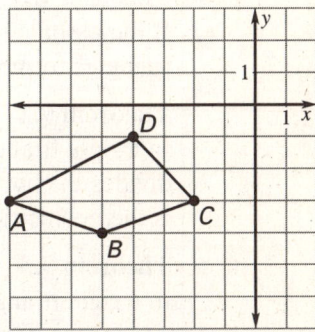

7. $k = \dfrac{1}{2}$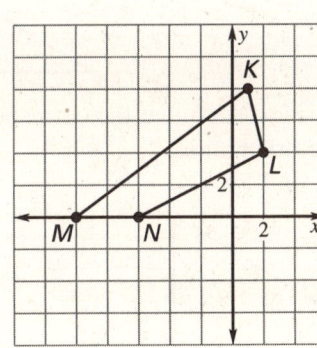

8. $k = \dfrac{3}{4}$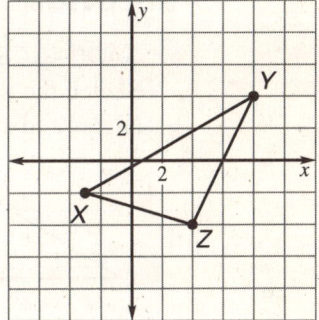

LESSON 9.1

Practice with Examples
For use with pages 527–534

GOAL Solve problems involving similar right triangles formed by the altitude drawn to the hypotenuse of a right triangle and use a geometric mean to solve problems

> **Theorem 9.1**
> If the altitude is drawn to the hypotenuse of a right triangle, then the two triangles formed are similar to the original triangle and to each other.
>
> **Theorem 9.2**
> In a right triangle, the altitude from the right angle to the hypotenuse divides the hypotenuse into two segments. The length of the altitude is the geometric mean of the lengths of the two segments.
>
> **Theorem 9.3**
> In a right triangle, the altitude from the right angle to the hypotenuse divides the hypotenuse into two segments. The length of each leg of the right triangle is a geometric mean of the lengths of the hypotenuse and the segment of the hypotenuse that is adjacent to the leg.

EXAMPLE 1 Finding the Height of a Triangle

Consider the right triangle shown.

a. Identify the similar triangles.

b. Find the height h of $\triangle ABC$.

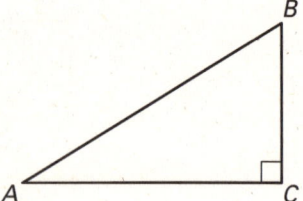

SOLUTION

a.  $\triangle ABC \sim \triangle CBD \sim \triangle ACD$

Sketch the three similar right triangles so that the corresponding angles and sides have the same orientation.

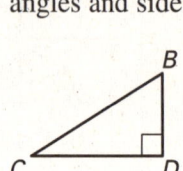 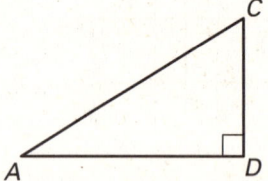

b. Use the fact that $\triangle ABC \sim \triangle CBD$ to write a proportion.

$\dfrac{CD}{AC} = \dfrac{CB}{AB}$ Corresponding side lengths are in proportion.

$\dfrac{h}{10} = \dfrac{6}{11.6}$ Substitute.

$11.6h = 6(10)$ Cross product property

$h \approx 5.2$ Solve for h.

LESSON 9.1 CONTINUED

NAME _____ DATE _____

Practice with Examples
For use with pages 527–534

Exercises for Example 1

Find the height, h, of the given right triangle.

1.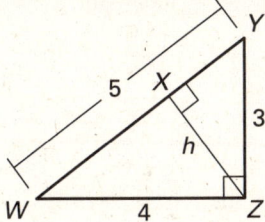
2.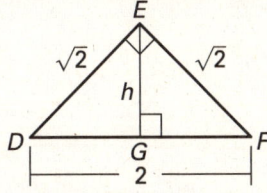
3.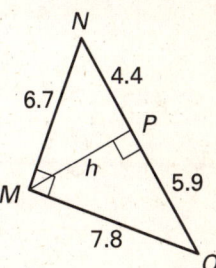

EXAMPLE 2 Using a Geometric Mean

Find the value of each variable.

a.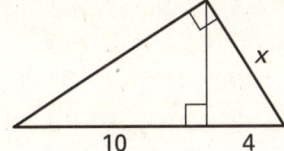

b.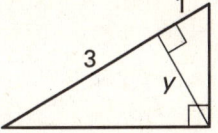

SOLUTION

a. Apply Theorem 9.3.

$$\frac{10 + 4}{x} = \frac{x}{4}$$

$$x^2 = 56$$

$$x = \sqrt{56} = \sqrt{4 \cdot 14} = 2\sqrt{14}$$

b. Apply Theorem 9.2.

$$\frac{3}{y} = \frac{y}{1}$$

$$y^2 = 3$$

$$y = \sqrt{3}$$

LESSON 9.1 CONTINUED

Practice with Examples
For use with pages 527–534

Exercises for Example 2
Find the value of each variable to the nearest tenth.

4.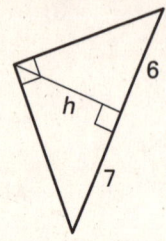

5.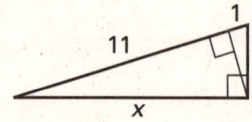

6.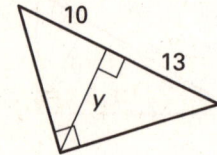

LESSON 9.2

NAME _____ DATE _____

Practice with Examples
For use with pages 535–541

GOAL Use the Pythagorean Theorem to solve problems

VOCABULARY

A **Pythagorean triple** is a set of three positive integers a, b, and c, that satisfy the equation $c^2 = a^2 + b^2$.

Theorem 9.4 Pythagorean Theorem
In a right triangle, the square of the length of the hypotenuse is equal to the sum of the squares of the lengths of the legs.

EXAMPLE 1 Find the Length of a Hypotenuse

Find the length of the hypotenuse of the right triangle.

Tell whether the side lengths form a Pythagorean triple.

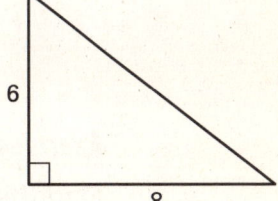

SOLUTION

(hypotenuse)^2 = (leg)^2 + (leg)^2	Pythagorean Theorem
$x^2 = 6^2 + 8^2$	Substitute.
$x^2 = 36 + 64$	Multiply.
$x^2 = 100$	Add.
$x = 10$	Find the positive square root.

Because the side lengths 6, 8, and 10 are integers, they form a Pythagorean triple.

Exercises for Example 1

Find the length of the hypotenuse of the right triangle. Tell whether the side lengths form a Pythagorean triple.

1.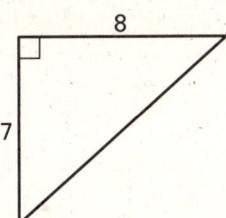

2.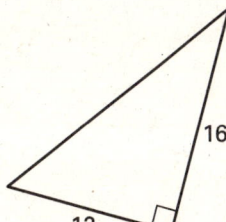

3.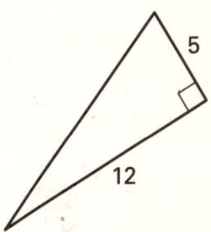

LESSON 9.2 CONTINUED

NAME _____ DATE _____

Practice with Examples

For use with pages 535–541

EXAMPLE 2 Finding the Length of a Leg

Find the length of the leg of the right triangle.

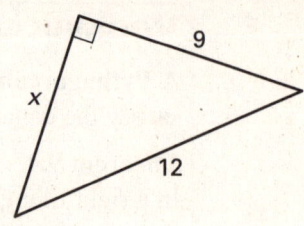

SOLUTION

$(\text{hypotenuse})^2 = (\text{leg})^2 + (\text{leg})^2$	Pythagorean Theorem
$12^2 = 9^2 + x^2$	Substitute.
$144 = 81 + x^2$	Multiply.
$63 = x^2$	Subtract 81 from each side.
$\sqrt{63} = x$	Find the positive square root.

Exercises for Example 2

Find the unknown side length. Round to the nearest tenth, if necessary.

4.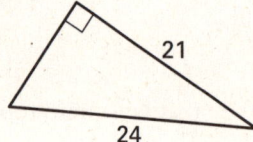

5.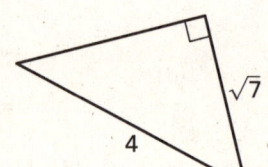

6.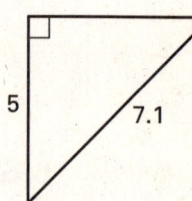

LESSON 9.2 CONTINUED

Practice with Examples
For use with pages 535–541

EXAMPLE 3 Finding the Area of a Triangle

Find the area of the triangle to the nearest tenth.

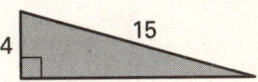

SOLUTION

In this case, the side of length 4 can be used as the height and the side of unknown length can be used as the base. To find the length of the unknown side, use the Pythagorean Theorem.

$(\text{hypotenuse})^2 = (\text{leg})^2 + (\text{leg})^2$ Pythagorean Theorem

$15^2 = 4^2 + b^2$ Substitute.

$\sqrt{209} = b$ Solve for b.

Now find the area of the triangle.

$A = \frac{1}{2}bh = \frac{1}{2}(\sqrt{209})(4) \approx 28.9$ square units

Exercises for Example 3

Find the area of the triangle to the nearest tenth.

7.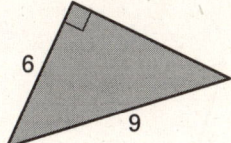

8.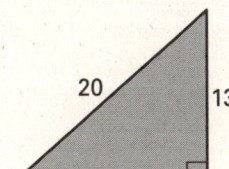

9.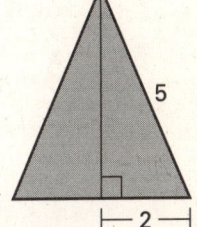

LESSON 9.3

Practice with Examples
For use with pages 543–548

GOAL Use the converse of the Pythagorean Theorem to solve problems and use side lengths to classify triangles by their angle measures

> **Theorem 9.5 Converse of the Pythagorean Theorem**
> If the square of the length of the longest side of a triangle is equal to the sum of the squares of the lengths of the other two sides, then the triangle is a right triangle.
>
> **Theorem 9.6**
> If the square of the length of the longest side of a triangle is less than the sum of the squares of the lengths of the other two sides, then the triangle is acute.
>
> **Theorem 9.7**
> If the square of the length of the longest side of a triangle is greater than the sum of the squares of the lengths of the other two sides, then the triangle is obtuse.

EXAMPLE 1   Verifying Right Triangles

The triangles below appear to be right triangles. Tell whether they are right triangles.

a. 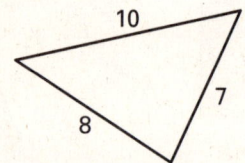   b.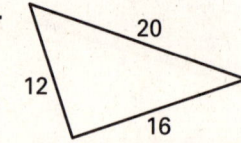

SOLUTION

Let c represent the length of the longest side of the triangle (you do not want to call this the "hypotenuse" because you do not yet know if the triangle is a right triangle). Check to see whether the side lengths satisfy the equation $c^2 = a^2 + b^2$.

a. $10^2 \stackrel{?}{=} 8^2 + 7^2$
    $100 \stackrel{?}{=} 64 + 49$
    $100 \neq 113$

The triangle is not a right triangle.

b. $20^2 \stackrel{?}{=} 12^2 + 16^2$
    $400 \stackrel{?}{=} 144 + 256$
    $400 = 400$

The triangle is a right triangle.

LESSON 9.3 CONTINUED

Practice with Examples

For use with pages 543–548

Exercises for Example 1

In Exercises 1–3, determine if the triangles are right triangles.

1.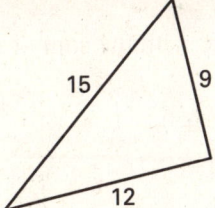
2.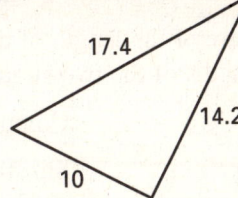
3.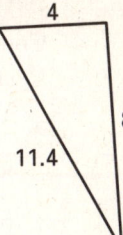

EXAMPLE 2 Classifying Triangles

Decide whether the set of numbers can represent the side lengths of a triangle. If they can, classify the triangle as *right*, *acute*, or *obtuse*.

a. 58, 69, 80

b. 11, 30, 39

LESSON 9.3 CONTINUED

Practice with Examples
For use with pages 543–548

SOLUTION

You can use the Triangle Inequality to confirm that each set of numbers can represent the side lengths of a triangle.

Compare the square of the length of the longest side with the sum of the squares of the lengths of the two shorter sides.

a. $c^2 \; ? \; a^2 + b^2$ Compare c^2 with $a^2 + b^2$.
   $80^2 \; ? \; 58^2 + 69^2$ Substitute.
   $6400 \; ? \; 3364 + 4761$ Multiply.
   $6400 < 8125$ c^2 is less than $a^2 + b^2$.
   Because $c^2 < a^2 + b^2$, the triangle is acute.

b. $c^2 \; ? \; a^2 + b^2$ Compare c^2 with $a^2 + b^2$.
   $39^2 \; ? \; 11^2 + 30^2$ Substitute.
   $1521 \; ? \; 121 + 900$ Multiply.
   $1521 > 1021$ c^2 is greater than $a^2 + b^2$.
   Because $c^2 > a^2 + b^2$, the triangle is obtuse.

Exercises for Example 2

Decide whether the set of numbers can represent the side lengths of a triangle. If they can, classify the triangle as *right*, *acute*, or *obtuse*.

4. 5, $\sqrt{56}$, 9   **5.** 23, 44, 70   **6.** 12, 80, 87   **7.** 4, 7, 10

LESSON 9.4

Practice with Examples

For use with pages 551–556

GOAL Find the side lengths of special right triangles

VOCABULARY

Right triangles whose angle measures are 45°- 45°- 90° or 30°- 60°- 90° are called **special right triangles.**

Theorem 9.8 The 45°- 45°- 90° Triangle Theorem
In a 45°- 45°- 90° triangle, the hypotenuse is $\sqrt{2}$ times as long as each leg.

Theorem 9.9 The 30°- 60°- 90° Triangle Theorem
In a 30°- 60°- 90° triangle, the hypotenuse is twice as long as the shorter leg, and the longer leg is $\sqrt{3}$ times as long as the shorter leg.

EXAMPLE 1 Finding Side Lengths in a 45°- 45°- 90° Triangle

Find the value of x.

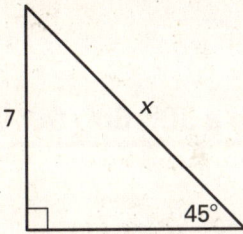

SOLUTION

By the Triangle Sum Theorem, the measure of the third angle is 45°. The triangle is a 45°- 45°- 90° right triangle, so the length x of the hypotenuse is $\sqrt{2}$ times the length of a leg.

$\text{Hypotenuse} = \sqrt{2} \cdot \text{leg}$ 45°- 45°- 90° Triangle Theorem

$x = \sqrt{2} \cdot 7$ Substitute.

$x = 7\sqrt{2}$ Simplify.

LESSON 9.4 CONTINUED

Practice with Examples

For use with pages 551–556

Exercises for Example 1

Find the value of each variable.

1.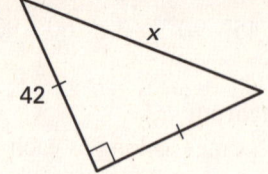
2.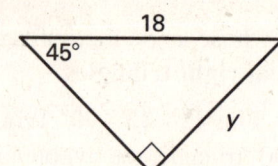
3.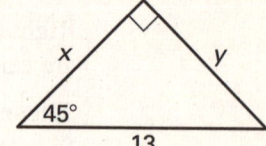

EXAMPLE 2 Finding Side Lengths in a 30°- 60°- 90° Triangle

Find the value of x.

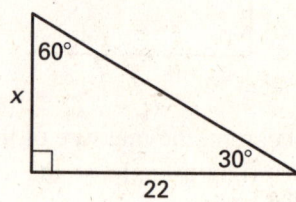

SOLUTION

Because the triangle is a 30°- 60°- 90° triangle, the longer leg is $\sqrt{3}$ times the length x of the shorter leg.

Longer leg = $\sqrt{3}$ · shorter leg	30°- 60°- 90° Triangle Theorem
$22 = \sqrt{3} \cdot x$	Substitute.
$\dfrac{22}{\sqrt{3}} = x$	Divide each side by $\sqrt{3}$.
$\dfrac{\sqrt{3}}{\sqrt{3}} \cdot \dfrac{22}{\sqrt{3}} = x$	Multiply numerator and denominator by $\sqrt{3}$.
$\dfrac{22\sqrt{3}}{3} = x$	Simplify.

LESSON 9.4 CONTINUED

NAME _____ DATE _____

Practice with Examples

For use with pages 551–556

Exercises for Example 2

Find the value of each variable.

4.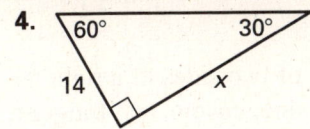

5.

6.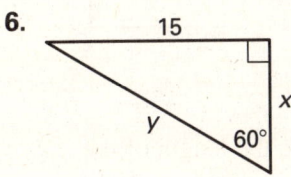

LESSON 9.5

Practice with Examples
For use with pages 558–565

GOAL Find the sine, the cosine, and the tangent of an acute angle and use trigonometric ratios to solve real-life problems

VOCABULARY

A **trigonometric ratio** is a ratio of the lengths of two sides of a right triangle. The three basic trigonometric ratios are **sine, cosine,** and **tangent,** which are abbreviated as *sin, cos,* and *tan,* respectively.

The angle that your line of sight makes with a line drawn horizontally is called the **angle of elevation.**

Trigonometric Ratios

Let $\triangle ABC$ be a right triangle. The sine, the cosine, and the tangent of the acute angle $\angle A$ are defined as follows.

$$\sin A = \frac{\text{side opposite } \angle A}{\text{hypotenuse}} = \frac{a}{c}$$

$$\cos A = \frac{\text{side adjacent } \angle A}{\text{hypotenuse}} = \frac{b}{c}$$

$$\tan A = \frac{\text{side opposite } \angle A}{\text{side adjacent } \angle A} = \frac{a}{b}$$

EXAMPLE 1 Finding Trigonometric Ratios

Find the sine, the cosine, and the tangent of the indicated angle.

a. $\angle A$

b. $\angle B$

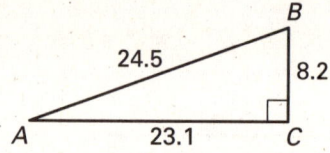

SOLUTION

a. The length of the hypotenuse is 24.5. For $\angle A$, the length of the opposite side is 8.2, and the length of the adjacent side is 23.1.

$$\sin A = \frac{\text{opp.}}{\text{hyp.}} = \frac{8.2}{24.5} \approx 0.3347$$

$$\cos A = \frac{\text{adj.}}{\text{hyp.}} = \frac{23.1}{24.5} \approx 0.9429$$

$$\tan A = \frac{\text{opp.}}{\text{adj.}} = \frac{8.2}{23.1} \approx 0.3550$$

LESSON 9.5 CONTINUED

Practice with Examples

For use with pages 558–565

b. The length of the hypotenuse is 24.5. For $\angle B$, the length of the opposite side is 23.1 and the length of the adjacent side is 8.2.

$$\sin B = \frac{\text{opp.}}{\text{hyp.}} = \frac{23.1}{24.5} \approx 0.9429$$

$$\cos B = \frac{\text{adj.}}{\text{hyp.}} = \frac{8.2}{24.5} \approx 0.3347$$

$$\tan B = \frac{\text{opp.}}{\text{adj.}} = \frac{23.1}{8.2} \approx 2.8171$$

Exercises for Example 1

Find the sine, cosine, and tangent of $\angle A$.

1.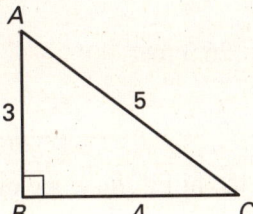

2.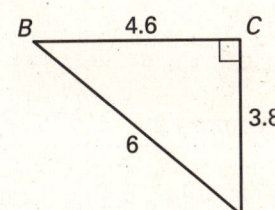

3.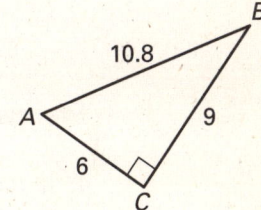

LESSON 9.5 CONTINUED

Practice with Examples

For use with pages 558–565

EXAMPLE 2 Estimating a Distance

It is known that a hill frequently used for sled riding has an angle of elevation of 30° at its bottom. If the length of a sledder's ride is 52.6 feet, estimate the height of the hill.

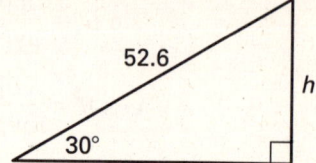

SOLUTION

Use the sine ratio for the 30° angle, because you have the value of the hypotenuse and you are looking for the value of the side opposite the 30° angle.

$$\sin 30° = \frac{h}{52.6}$$

$$h = (52.6) \cdot \sin 30° = (52.6) \cdot (0.5) = 26.3 \text{ feet}$$

Exercises for Example 2

4. In the sled-riding example, find the height of the hill if the angle of elevation of the hill is 42°.

5. If the angle of elevation from your position on the ground to the top of a building is 67° and you are standing 30 meters from the foot of the building, approximate the height of the building.

LESSON 9.6

NAME _____ DATE _____

Practice with Examples

For use with pages 567–572

GOAL Solve a right triangle

VOCABULARY

To **solve a right triangle** means to determine the measures of all six parts (the right angle, the two acute angles, the hypotenuse, and the two legs).

EXAMPLE 1 *Solving a Right Triangle*

Solve the right triangle.

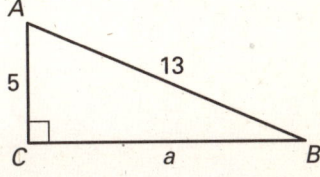

SOLUTION

Begin by using the Pythagorean Theorem to find the length of the missing side.

$(\text{hypotenuse})^2 = (\text{leg})^2 + (\text{leg})^2$ Pythagorean Theorem

$13^2 = a^2 + 5^2$ Substitute.

$169 = a^2 + 25$ Multiply.

$144 = a^2$ Subtract 25 from each side.

$12 = a$ Find the positive square root.

Then find the measure of $\angle B$.

$\tan B = \dfrac{\text{opp.}}{\text{adj.}}$

$\tan B = \dfrac{5}{12}$ Substitute.

$m\angle B \approx 22.6°$ Use a calculator.

Finally, because $\angle A$ and $\angle B$ are complements, you can write $m\angle A = 90° - m\angle B \approx 90° - 22.6° = 67.4°$.

The side lengths of $\triangle ABC$ are 5, 12, and 13. $\triangle ABC$ has one right angle and two acute angles whose measures are about 22.6° and 67.4°.

LESSON 9.6 CONTINUED

Practice with Examples
For use with pages 567–572

Exercises for Example 1
Solve the right triangle.

1.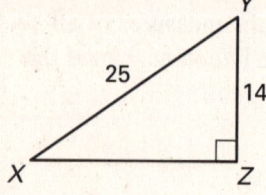
2.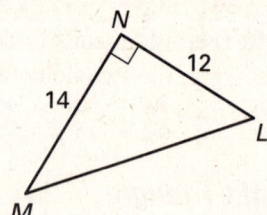
3.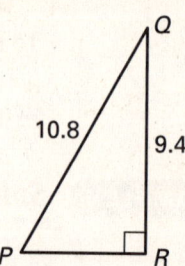

EXAMPLE 2 Solving a Right Triangle

Solve the right triangle.

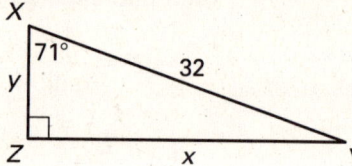

SOLUTION

Use trigonometric ratios to find the values of x and y.

$\sin X = \dfrac{\text{opp.}}{\text{hyp.}}$ $\qquad\qquad$ $\cos X = \dfrac{\text{adj.}}{\text{hyp.}}$

$\sin 71° = \dfrac{x}{32}$ $\qquad\qquad$ $\cos 71° = \dfrac{y}{32}$

$32 \sin 71° = x$ $\qquad\qquad$ $32 \cos 71° = y$

$32(0.9455) = x$ $\qquad\qquad$ $32(0.3256) = y$

$30.3 \approx x$ $\qquad\qquad$ $10.4 \approx y$

Because $\angle X$ and $\angle Y$ are complements, you can write

$m\angle Y = 90° - m\angle x = 90° - 71° = 19°$.

The side lengths of the triangle are about 10.4, 30.3, and 32. The triangle has one right angle and two acute angles whose measures are 71° and 19°.

LESSON 9.6 CONTINUED

Practice with Examples
For use with pages 567–572

Exercises for Example 2

Solve the right triangle.

4.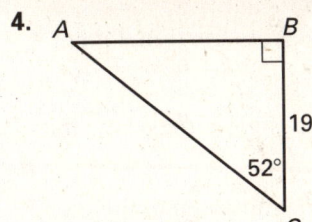

5.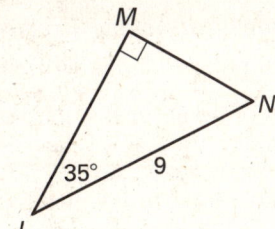

6.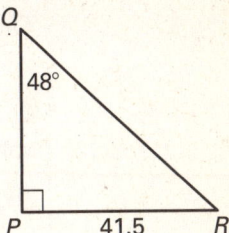

LESSON 9.7

Practice with Examples

For use with pages 573–579

GOAL Find the magnitude and the direction of a vector and add vectors

VOCABULARY

The **magnitude of a vector** $\overrightarrow{AB}$ is the distance from the initial point A to the terminal point B and is written $|\overrightarrow{AB}|$.

The **direction of a vector** is determined by the angle it makes with a horizontal line.

Two vectors are **equal** if they have the same magnitude and direction.

Two vectors are **parallel** if they have the same or opposite directions.

Sum of Two Vectors

The sum of $\vec{u} = \langle a_1, b_1 \rangle$ and $\vec{v} = \langle a_2, b_2 \rangle$ is
$\vec{u} + \vec{v} = \langle a_1 + a_2, b_1 + b_2 \rangle$.

EXAMPLE 1 Finding the Magnitude of a Vector

Points P and Q are the initial and terminal points of the vector $\overrightarrow{PQ}$.
Draw $\overrightarrow{PQ}$ in a coordinate plane. Write the component form of the vector and find its magnitude.

a. $P(1, 2), Q(5, 5)$ **b.** $P(0, 4), Q(-2, -4)$

SOLUTION

a. Component form = $\langle x_2 - x_1, y_2 - y_1 \rangle$

$\overrightarrow{PQ} = \langle 5 - 1, 5 - 2 \rangle$
$= \langle 4, 3 \rangle$

Use the Distance Formula to find the magnitude.

$|\overrightarrow{PQ}| = \sqrt{(5-1)^2 + (5-2)^2} = \sqrt{25} = 5$

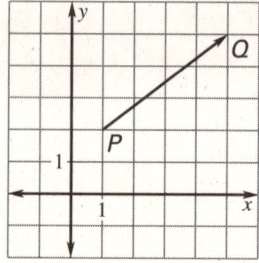

b. Component form = $\langle x_2 - x_1, y_2 - y_1 \rangle$

$\overrightarrow{PQ} = \langle -2 - 0, -4 - 4 \rangle$
$= \langle -2, -8 \rangle$

Use the Distance Formula to find the magnitude.

$|\overrightarrow{PQ}| = \sqrt{(-2-0)^2 + (-4-4)^2} = \sqrt{68} \approx 8.2$

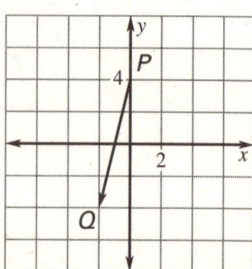

LESSON 9.7 CONTINUED

Practice with Examples

For use with pages 573–579

Exercises for Example 1

Draw $\vec{PQ}$ in a coordinate plane. Write the component form of the vector and find its magnitude.

1. $P(3, 2), Q(1, 9)$

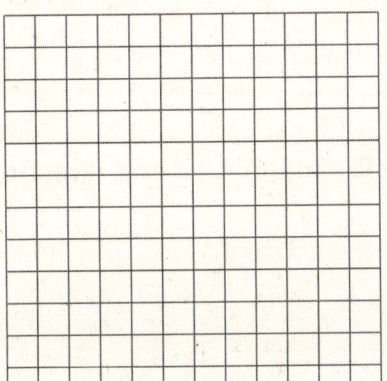

2. $P(-2, 1), Q(0, -5)$

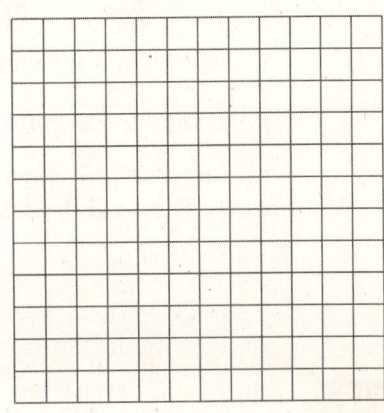

3. $P(3, 8), Q(-1, 10)$

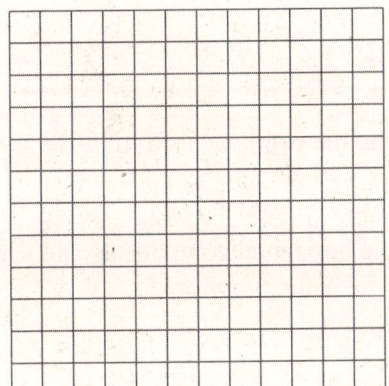

4. $P(-4, -11), Q(0, 2)$

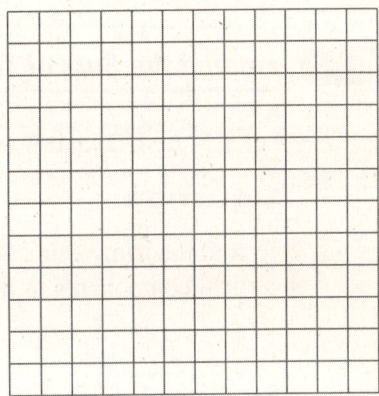

EXAMPLE 2 Describing the Direction of a Vector

The vector $\vec{AB}$ depicts the velocity of a moving vehicle. The scale on each axis is in kilometers per hour. Find the (a) speed of the vehicle and (b) direction it is traveling relative to east.

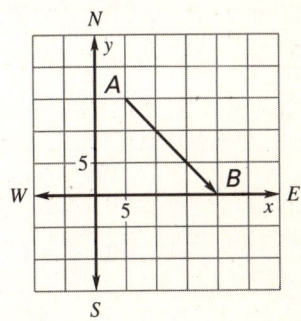

SOLUTION

a. The magnitude of the vector $\vec{AB}$ represents the vehicle's speed. Use the Distance Formula.

$|\vec{AB}| = \sqrt{(20-5)^2 + (0-15)^2}$

$= \sqrt{15^2 + 15^2}$

≈ 21.2

The speed of the vehicle is about 21.2 kilometers per hour.

LESSON 9.7 CONTINUED

Practice with Examples
For use with pages 573–580

b. The tangent of the angle formed by the vector and a line drawn parallel to the x-axis at point A is $-\frac{15}{15} = -1$. Use a calculator to find the angle measure.

-1 [2nd] [TAN] $\approx$ $-45°$

The vehicle is traveling in a direction 45° south of east.

Exercises for Example 2

In Exercises 5–7, find the vehicle's magnitude and direction if points A and B are as given.

5. $A(0, 0)$, $B(6, 7)$
6. $A(-2, 4)$, $B(3, -1)$
7. $A(2, 4)$, $B(-3, -1)$

EXAMPLE 3 Finding the Sum of Two Vectors

Let $\vec{u} = \langle -4, 2 \rangle$ and $\vec{v} = \langle 3, 1 \rangle$. Write the component form of the sum $\vec{u} + \vec{v}$.

SOLUTION

To find the sum vector $\vec{u} + \vec{v}$, add the horizontal components and add the vertical components of $\vec{u}$ and $\vec{v}$.

$\vec{u} + \vec{v} = \langle -4 + 3, 2 + 1 \rangle = \langle -1, 3 \rangle$

Exercises for Example 3

For the given vectors $\vec{u}$ and $\vec{v}$, find the component form of the sum $\vec{u} + \vec{v}$.

8. $\vec{u} = \langle 0, 8 \rangle$ and $\vec{v} = \langle -3, 5 \rangle$
9. $\vec{u} = \langle -2, -7 \rangle$ and $\vec{v} = \langle 2, 10 \rangle$

10. $\vec{u} = \langle 3, 12 \rangle$ and $\vec{v} = \langle -3, -12 \rangle$

LESSON 10.1

Practice with Examples
For use with pages 595–602

GOAL Identify segments and lines related to circles and use properties of a tangent to a circle

VOCABULARY

A **circle** is the set of all points in a plane that are equidistant from a given point, called the **center** of the circle.

The distance from the center to a point on the circle is the **radius** of the circle.

Two circles are **congruent** if they have the same radius.

The distance across the circle, through its center, is the **diameter** of the circle.

A **radius** is a segment whose endpoints are the center of the circle and a point on the circle.

A **chord** is a segment whose endpoints are points on the circle.

A **diameter** is a chord that passes through the center of the circle.

A **secant** is a line that intersects a circle in two points.

A **tangent** is a line in the plane of a circle that intersects the circle in exactly one point.

Theorem 10.1
If a line is tangent to a circle, then it is perpendicular to the radius drawn to the point of tangency.

Theorem 10.2
In a plane, if a line is perpendicular to a radius of a circle at its endpoint on the circle, then the line is tangent to the circle.

Theorem 10.3
If two segments from the same exterior point are tangent to a circle, then they are congruent.

EXAMPLE 1 *Identifying Special Segments and Lines*

Tell whether the line or segment is best described as a *chord*, a *secant*, a *tangent*, a *diameter*, or a *radius* of ⊙C.

a. $\overline{HC}$
b. $\overleftrightarrow{DG}$
c. $\overline{BE}$
d. $\overleftrightarrow{AF}$
e. $\overline{BH}$

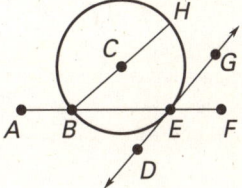

LESSON 10.1 CONTINUED

Practice with Examples
For use with pages 595–602

SOLUTION

a. $\overline{HC}$ is a radius because one of its endpoints, *H*, is a point on the circle and the other endpoint, *C*, is the circle's center.

b. $\overleftrightarrow{DG}$ is a tangent because it intersects the circle in one point.

c. $\overline{BE}$ is a chord because its endpoints are on the circle.

d. $\overleftrightarrow{AF}$ is a secant because it intersects the circle in two points.

e. $\overline{BH}$ is a diameter because its endpoints are on the circle and it contains the center *C*.

Exercises for Example 1

In Exercises 1–8, tell whether the line or segment is best described as a *chord*, a *secant*, a *tangent*, a *diameter*, or a *radius* of ⊙*C*.

1. $\overline{AB}$
2. $\overleftrightarrow{DE}$

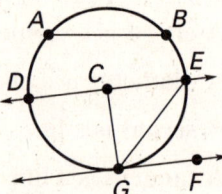

3. $\overline{DC}$
4. $\overline{DE}$

5. $\overleftrightarrow{FG}$
6. $\overline{CG}$

7. $\overline{EG}$
8. $\overline{EC}$

LESSON 10.1 CONTINUED

Practice with Examples

For use with pages 595–602

EXAMPLE 2 Using Properties of Tangents

$\overleftrightarrow{AB}$ and $\overleftrightarrow{AD}$ are tangent to $\odot C$. Find the value of x.

SOLUTION

$AB = AD$	Two tangent segments from the same point are congruent.
$6x - 3 = 5x + 7$	Substitute.
$6x = 5x + 10$	Add 3 to each side.
$x = 10$	Subtract $5x$ from each side.

Exercises for Example 2

In Exercises 9–11, $\overleftrightarrow{AB}$ and $\overleftrightarrow{AD}$ are tangent to $\odot C$. Find the value of x.

9.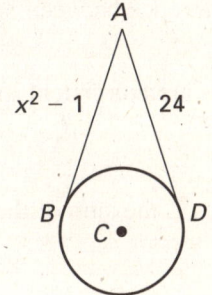

10.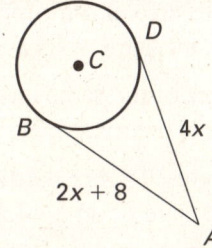

11.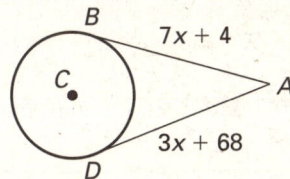

LESSON 10.2

Practice with Examples
For use with pages 603–611

GOAL Use properties of arcs of circles and use properties of chords of circles

VOCABULARY

In a plane, an angle whose vertex is the center of a circle is a **central angle** of the circle.

If the measure of a central angle, $\angle APB$, is less than 180°, then A and B and the points of $\odot P$ in the interior of $\angle APB$ form a **minor arc** of the circle.

The points A and B and the points of $\odot P$ in the *exterior* of $\angle APB$ form a **major arc** of the circle. If the endpoints of an arc are the endpoints of a diameter, then the arc is a **semicircle.**

The **measure of a minor arc** is defined to be the measure of its central angle.

The **measure of a major arc** is defined as the difference between 360° and the measure of its associated minor arc.

Two arcs of the same circle or of congruent circles are **congruent arcs** if they have the same measure.

Postulate 26   Arc Addition Postulate
The measure of an arc formed by two adjacent arcs is the sum of the measures of the two arcs.

Theorem 10.4
In the same circle, or in congruent circles, two minor arcs are congruent if and only if their corresponding chords are congruent.

Theorem 10.5
If a diameter of a circle is perpendicular to a chord, then the diameter bisects the chord and its arc.

Theorem 10.6
If one chord is a perpendicular bisector of another chord, then the first chord is a diameter.

Theorem 10.7
In the same circle, or in congruent circles, two chords are congruent if and only if they are equidistant from the center.

LESSON 10.2 CONTINUED

Practice with Examples

For use with pages 603–611

EXAMPLE 1 Finding Measures of Arcs

Find the measure of each arc of ⊙C.

a. $\overset{\frown}{AD}$ b. $\overset{\frown}{ADB}$
c. $\overset{\frown}{DBA}$ d. $\overset{\frown}{BD}$

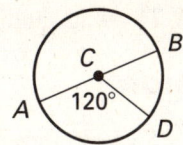

SOLUTION

a. $\overset{\frown}{AD}$ is a minor arc, so $m\overset{\frown}{AD} = m\angle ACD = 120°$.

b. $\overset{\frown}{ADB}$ is a semicircle, so $m\overset{\frown}{ADB} = 180°$.

c. $\overset{\frown}{DBA}$ is a major arc, so $m\overset{\frown}{DBA} = 360° - 120° = 240°$.

d. $\overset{\frown}{BD}$ is a minor arc, so $m\overset{\frown}{BD} = m\angle BCD$. Because $\angle BCD$ and $\angle ACD$ form a linear pair, $m\angle BCD = 180° - m\angle ACD = 180° - 120° = 60°$. So $m\overset{\frown}{BD} = 60°$.

Exercises for Example 1

Find the measure of the arcs of the given circle.

1. Find the measure of each arc of ⊙C.

   a. $\overset{\frown}{ADB}$ b. $\overset{\frown}{AD}$

   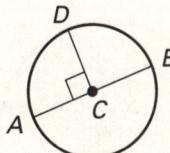

   c. $\overset{\frown}{DB}$ d. $\overset{\frown}{DBA}$

2. Find the measure of each arc of ⊙Q.

   a. $\overset{\frown}{PR}$ b. $\overset{\frown}{PRS}$

   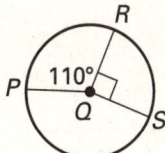

   c. $\overset{\frown}{PS}$ d. $\overset{\frown}{RSP}$

LESSON 10.2 CONTINUED

Practice with Examples

For use with pages 603–611

EXAMPLE 2 Using Theorem 10.7

$PS = 12$, $TV = 12$, and $SQ = 7$. Find QU.

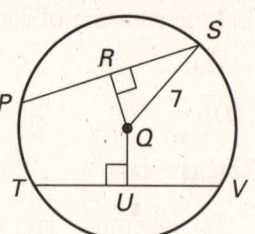

SOLUTION

Because $\overline{PS} \cong \overline{TV}$, they are equidistant from the center by Theorem 10.7. To find QU, first find QR. $\overline{QR} \perp \overline{PS}$, so $\overline{QR}$ bisects $\overline{PS}$. Because $PS = 12$, $RS = \dfrac{12}{2} = 6$. Now look at $\triangle QRS$ which is a right triangle. Use the Pythagorean Theorem to find QR. $QR = \sqrt{QS^2 - RS^2} = \sqrt{7^2 - 6^2} = \sqrt{13}$. Because $\overline{QR} \cong \overline{QU}$, $QR = QU = \sqrt{13}$.

Exercises for Example 2

Use the given information to find the value of *x*.

3. $AB = DE = 10$, radius $= 6$

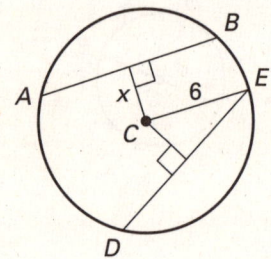

4. $QV = 2$, $QU = 2$, $SU = 3$

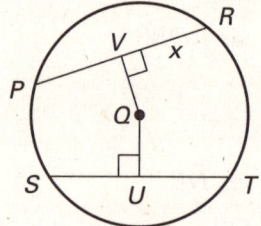

186 Geometry
Practice Workbook with Examples

LESSON 10.3

NAME _____ DATE _____

Practice with Examples

For use with pages 613–620

GOAL Use inscribed angles to solve problems and use properties of inscribed polygons

VOCABULARY

An **inscribed angle** is an angle whose vertex is on a circle and whose sides contain chords of the circle.

The arc that lies in the interior of an inscribed angle and has endpoints on the angle is called the **intercepted arc** of the angle.

If all of the vertices of a polygon lie on a circle, the polygon is **inscribed** in the circle and the circle is **circumscribed** about the polygon.

Theorem 10.8 Measure of an Inscribed Angle
If an angle is inscribed in a circle, then its measure is half the measure of its intercepted arc.

Theorem 10.9
If two inscribed angles of a circle intercept the same arc, then the angles are congruent.

Theorem 10.11
A quadrilateral can be inscribed in a circle if and only if its opposite angles are supplementary.

EXAMPLE 1 Finding Measures of Arcs and Inscribed Angles

Find the value of x.

a.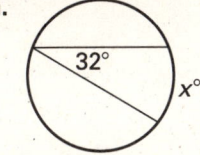

b.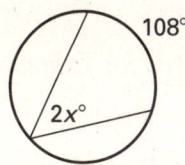

SOLUTION

a. By Theorem 10.8,

$$32° = \frac{1}{2}x°$$

$$64 = x$$

b. $2x° = \frac{1}{2}(108°)$

$2x = 54$

$x = 27$

Geometry
Practice Workbook with Examples

LESSON 10.3 CONTINUED

Practice with Examples
For use with pages 613–620

Exercises for Example 1

Find the value of x.

1.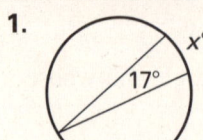
2.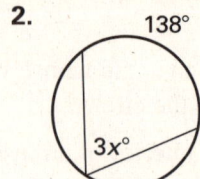
3.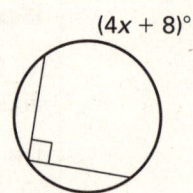

EXAMPLE 2 Finding the Measure of an Angle

If ∠CAD is a right angle, what is the measure of ∠CBD?

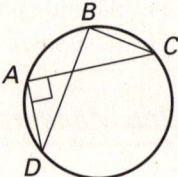

SOLUTION

By Theorem 10.9, ∠CAD ≅ ∠CBD because the two angles intercept the same arc. So, m∠CBD = 90°.

Exercises for Example 2

Find the value of x.

4.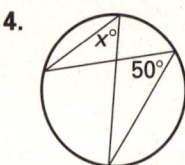
5.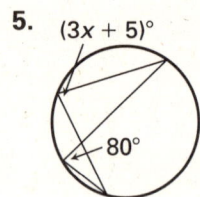
6.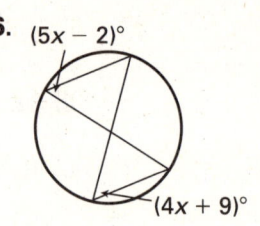

LESSON 10.3 CONTINUED

NAME _____ DATE _____

Practice with Examples
For use with pages 613–620

EXAMPLE 3 Using an Inscribed Quadrilateral

Find the value of each variable.

SOLUTION

By Theorem 10.11, the opposite angles of this quadrilateral are supplementary. So you can write the following equations and then solve for the variable in each.

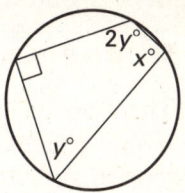

$x° + 90° = 180°$ $2y° + y° = 180°$
$x = 90$ $3y° = 180°$
                      $y = 60$

Exercises for Example 3

Find the value of each variable.

7.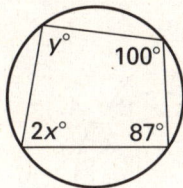

8.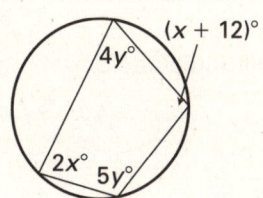

9.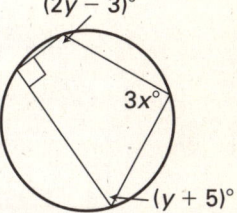

Geometry
Practice Workbook with Examples   189

LESSON 10.4

Practice with Examples

For use with pages 621–627

GOAL Use angles formed by tangents and chords to solve problems in geometry and use angles formed by lines that intersect a circle to solve problems

VOCABULARY

Theorem 10.12
If a tangent and a chord intersect at a point on a circle, then the measure of each angle formed is one half the measure of its intercepted arc.

Theorem 10.13
If two chords intersect in the *interior* of a circle, then the measure of each angle is one half the *sum* of the measures of the arcs intercepted by the angle and its vertical angle.

Theorem 10.14
If a tangent and a secant, two tangents, or two secants intersect in the *exterior* of a circle, then the measure of the angle formed is one half the *difference* of the measures of the intercepted arcs.

EXAMPLE 1 Finding Angle and Arc Measures

Line *m* is tangent to the circle.

a. Find $m\angle 1$

b. $m\widehat{ACB}$

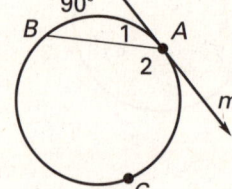

SOLUTION

a. $m\angle 1 = \frac{1}{2}(90°) = 45°$

b. Because $\angle 1$ and $\angle 2$ are a linear pair,
$m\angle 2 = 180° - m\angle 1 = 180° - 45° = 135°$. So,
$m\widehat{ACB} = 2(135°) = 270°$.

LESSON 10.4 CONTINUED

Practice with Examples
For use with pages 621–627

Exercises for Example 1

Find the value of each variable.

1.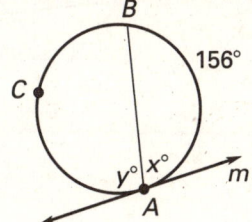
2.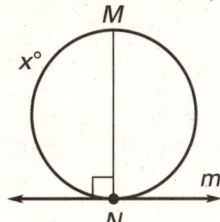
3.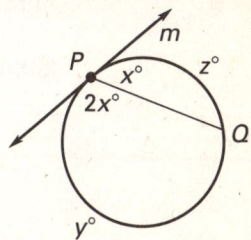

EXAMPLE 2 Using Theorem 10.13

Find the value of x.

SOLUTION

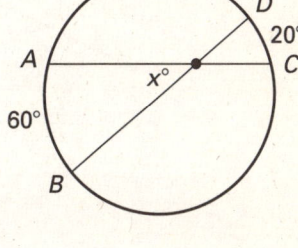

$x° = \frac{1}{2}(m\widehat{AB} + m\widehat{CD})$ Apply Theorem 10.13.

$x° = \frac{1}{2}(60° + 20°)$ Substitute.

$x = 40$ Simplify.

Exercises for Example 2

Find the value of x.

4.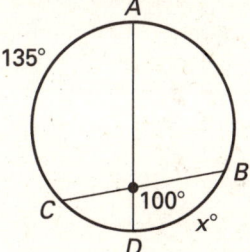
5.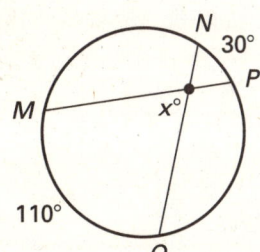
6.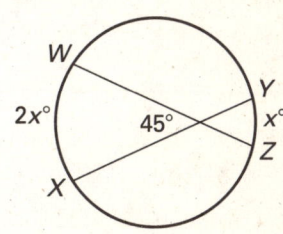

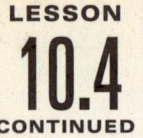

LESSON 10.4 CONTINUED

NAME _____ DATE _____

Practice with Examples
For use with pages 621–627

EXAMPLE 3 *Using Theorem 10.14*

Find the value of x.

SOLUTION

$x° = \frac{1}{2}(m\widehat{BC} - m\widehat{DE})$ Apply Theorem 10.14.

$x° = \frac{1}{2}(171° - 85°)$ Substitute.

$x = 43$ Simplify.

Exercises for Example 3

Find the value of x.

7.
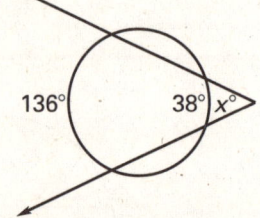

8.
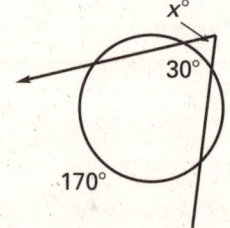

9.
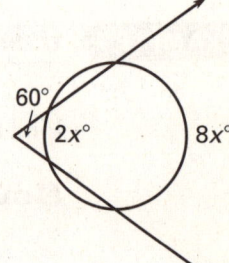

192 **Geometry**
Practice Workbook with Examples

LESSON 10.5

Practice with Examples
For use with pages 629–635

GOAL Find the lengths of segments of chords, tangents, and secants

VOCABULARY

In the figure shown, $\overline{PS}$ is a **tangent segment** because it is tangent to the circle at an endpoint. $\overline{PR}$ is a **secant segment** because one of the two intersection points with the circle is an endpoint. $\overline{PQ}$ is the **external segment** of $\overline{PR}$.

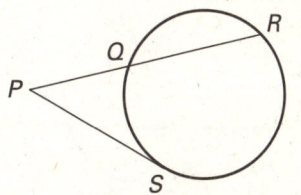

Theorem 10.15
If two chords intersect in the interior of a circle, then the product of the lengths of the segments of one chord is equal to the product of the lengths of the segments of the other chord.

Theorem 10.16
If two secant segments share the same endpoint outside a circle, then the product of the length of one secant segment and the length of its external segment equals the product of the length of the other secant segment and the length of its external segment.

Theorem 10.17
If a secant segment and a tangent segment share an endpoint outside a circle, then the product of the length of the secant segment and the length of its external segment equals the square of the length of the tangent segment.

EXAMPLE 1 *Finding Segment Lengths Using Theorem 10.15*

Find the value of x.

SOLUTION

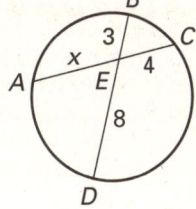

Because $\overline{AC}$ and $\overline{BD}$ are chords which intersect in the interior of the circle, Theorem 10.15 applies.

$EC \cdot EA = EB \cdot ED$ Use Theorem 10.15.

$4 \cdot x = 3 \cdot 8$ Substitute.

$4x = 24$ Simplify.

$x = 6$ Divide each side by 4.

LESSON 10.5 CONTINUED

Practice with Examples
For use with pages 629–635

Exercises for Example 1

Find the value of *x*.

1.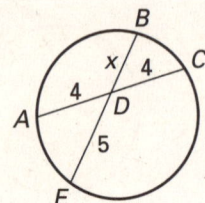

2.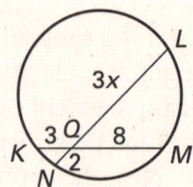

3.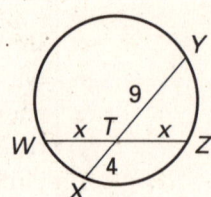

EXAMPLE 2 Finding Segment Lengths Using Theorem 10.16

Find the value of *x*.

SOLUTION

$CB \cdot CA = CE \cdot CD$	Use Theorem 10.16.
$4 \cdot (6 + 4) = 5 \cdot (x + 5)$	Substitute.
$40 = 5x + 25$	Simplify.
$15 = 5x$	Subtract 25 from each side.
$x = 3$	Divide each side by 5.

Exercises for Example 2

Find the value of *x*.

4.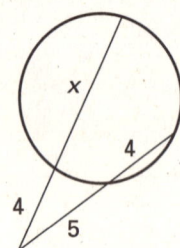

5.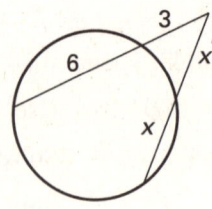

6.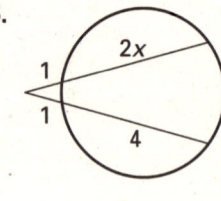

LESSON 10.5 CONTINUED

Practice with Examples
For use with pages 629–635

EXAMPLE 3 *Finding Segment Lengths Using Theorem 10.17*

Find the value of *x*.

SOLUTION

$CB \cdot CA = (CD)^2$ Use Theorem 10.17.

$4 \cdot (5 + 4) = x^2$ Substitute.

$36 = x^2$ Simplify.

$\pm 6 = x$ Take the square root of each side.

Use the positive solution, because lengths cannot be negative. So, $x = 6$.

Exercises for Example 3

Find the value of *x*.

7.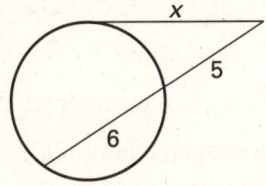

8.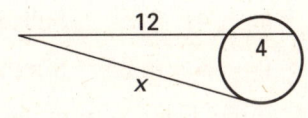

9.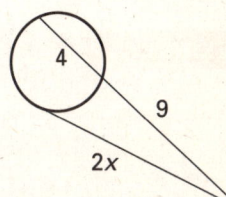

LESSON 10.6

Practice with Examples
For use with pages 636–640

GOAL Write the equation of a circle and use it and its graph to solve problems

VOCABULARY

The **standard equation of a circle** with radius r and center (h, k) is $(x - h)^2 + (y - k)^2 = r^2$.

EXAMPLE 1 *Writing a Standard Equation of a Circle*

a. Write the standard equation of the circle with center $(2, 4)$ and radius 5.

b. The point $(-2, 4)$ is on a circle whose center is $(0, 2)$. Write the standard equation of the circle.

SOLUTION

a. $(x - h)^2 + (y - k)^2 = r^2$ Standard equation of a circle
$(x - 2)^2 + (y - 4)^2 = 5^2$ Substitute.
$(x - 2)^2 + (y - 4)^2 = 25$ Simplify.

b. The radius is the distance from the point $(-2, 4)$ to the center $(0, 2)$.
$r = \sqrt{(0 - (-2))^2 + (2 - 4)^2}$ Use the Distance Formula.
$r = \sqrt{2^2 + (-2)^2} = 2\sqrt{2}$ Simplify.

Substitute $(h, k) = (0, 2)$ and $r = 2\sqrt{2}$ into the standard equation of a circle.
$(x - h)^2 + (y - k)^2 = r^2$ Standard equation of a circle
$(x - 0)^2 + (y - 2)^2 = (2\sqrt{2})^2$ Substitute.
$x^2 + (y - 2)^2 = 8$ Simplify.

Exercises for Example 1

Write the standard equation of the circle described.

1. center $(4, -1)$, radius 6

2. center $(-1, -5)$, radius 3.2

3. The center is $(-2, 3)$, a point on the circle is $(2, 3)$.

LESSON 10.6 CONTINUED

Practice with Examples

For use with pages 636–640

EXAMPLE 2 Graphing a Circle

The equation of a circle is $(x - 1)^2 + (y + 3)^2 = 25$. Graph the circle.

SOLUTION

Rewrite the equation to find the center and radius:

$(x - 1)^2 + (y + 3)^2 = 25$

$(x - 1)^2 + (y - (-3))^2 = 5^2$

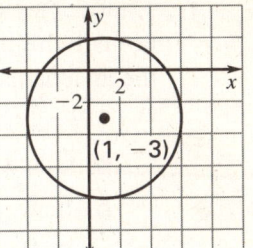

The center is $(1, -3)$ and the radius is 5. To graph the circle, place the point of a compass at $(1, -3)$, set the radius at 5 units, and swing the compass to draw a full circle.

Exercises for Example 2

Graph the circle that has the given equation.

4. $(x - 2)^2 + (y - 7)^2 = 4$

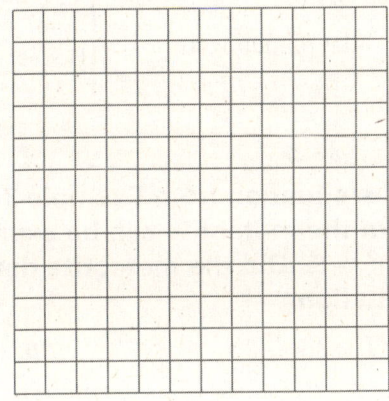

5. $(x + 6)^2 + (y - 4)^2 = 9$

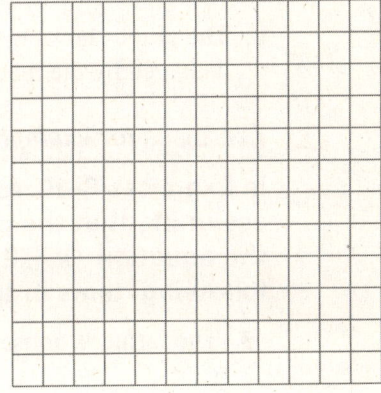

6. $(x + 3)^2 + y^2 = 16$

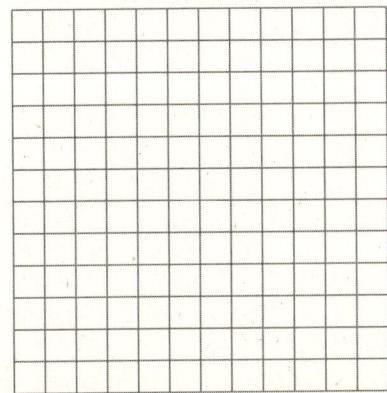

7. $x^2 + (y + 2)^2 = \dfrac{1}{2}$

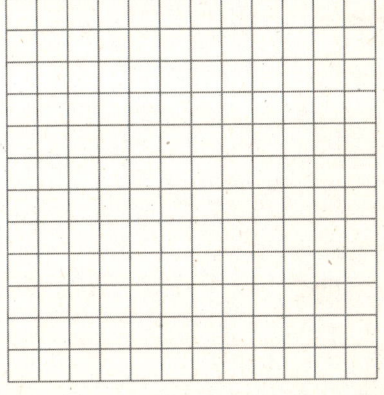

LESSON 10.6 CONTINUED

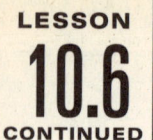

NAME _____ DATE _____

Practice with Examples

For use with pages 636–640

EXAMPLE 3 Applying Graphs of Circles

A farmer's plot of land was struck by a meteorite which damaged a circular area of his farm. If the farmer's house is labeled as the origin of a coordinate plane, the area damaged by the meteorite can be expressed by the equation $(x - 6)^2 + (y - 7)^2 = 16$.

  a. Graph the damaged area of the farm.

  b. Items on the farm are located as follows: A silo is at (2, 4), a barn is at (4, 6), and a pigpen is at (8, 9). Which of these items were damaged by the meteorite?

SOLUTION

  a. Rewrite the equation to find the center and radius:

  $(x - 6)^2 + (y - 7)^2 = 16$

  $(x - 6)^2 + (y - 7)^2 = 4^2$

  The center is (6, 7) and the radius is 4.

  b. The graph shows that the barn and the pigpen were damaged by the meteorite.

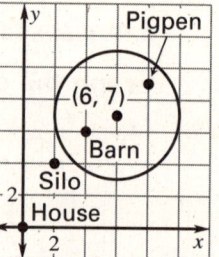

Exercises for Example 3

In Exercises 8–10, reconsider the situation from Example 3 above, assuming that the damage from the meteorite can be expressed by the equation $(x - 3)^2 + (y - 3)^2 = 9$. Did the meteorite damage the following items in this new situation?

  8. The farmer's house   **9.** The silo   **10.** The pigpen

LESSON 10.7

Practice with Examples
For use with pages 642–648

GOAL Draw the locus of points that satisfy a given condition and draw the locus of points that satisfy two or more conditions

VOCABULARY

A **locus** in a plane is the set of all points in a plane that satisfy a given condition or set of given conditions.

Finding a Locus
To find the locus of points that satisfy a given condition, use the following steps.

1. Draw any figures that are given in the statement of the problem.
2. Locate several points that satisfy the given condition.
3. Continue drawing points until you can recognize the pattern.
4. Draw the locus and describe it in words.

EXAMPLE 1 Finding a Locus

Sketch and describe the locus of points that satisfy the given condition(s)

a. in the interior of $\angle P$ and equidistant from both sides of $\angle P$.

b. equidistant from j and k.

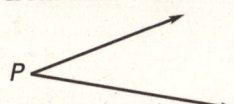
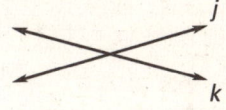

SOLUTION

a.

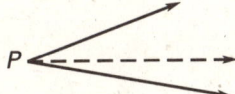

The points that make up the locus form the bisector of $\angle P$.

b.

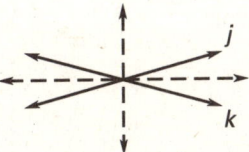

The points that make up the locus form the four angle bisectors of the angles formed by the intersection of the two lines.

LESSON 10.7 CONTINUED

Practice with Examples
For use with pages 642–648

Exercises for Example 1

Sketch and describe the locus of points that satisfy the given condition(s).

1. Equidistant from the lines $y = 3$ and $y = 7$

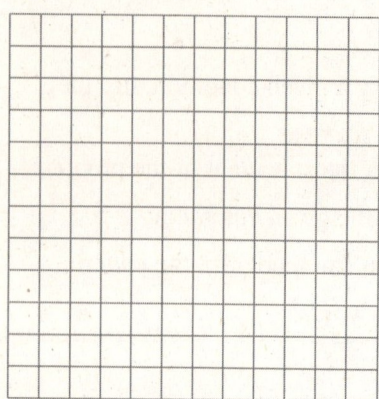

2. Within five units of the point $(-1, 2)$

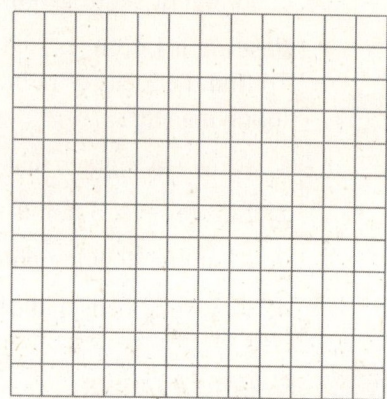

3. Equidistant from $A(-1, 1)$ and $B(1, -1)$

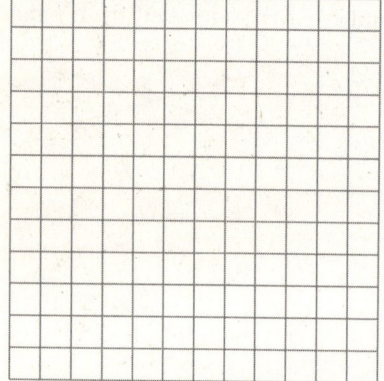

EXAMPLE 2 Drawing a Locus Satisfying Two Conditions

Sketch and describe the locus of points in the plane that are equidistant from A and B and less than 3 units from the origin.

LESSON 10.7 CONTINUED

Practice with Examples
For use with pages 642–648

SOLUTION

First, find the locus of points that are equidistant from A and B. This is the line $x = 0$.

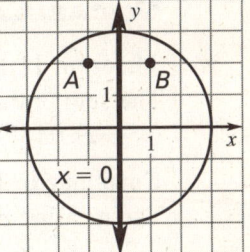

Next, find the locus of points that are less than 3 units from the origin. This is the circle centered at the origin with radius of 3.

Now, find the overlap of these two loci. This is the line segment from point $(0, 3)$ to $(0, -3)$.

Exercises for Example 2

Sketch and describe the locus of points in the plane that satisfy the given conditions. Explain your reasoning.

4. Equidistant from A and B and 2 units from the point $(1, 1)$.

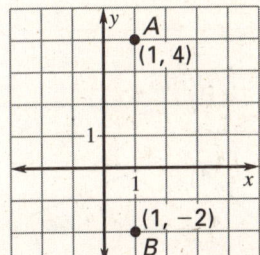

5. Equidistant from l and m and within four units from the origin.

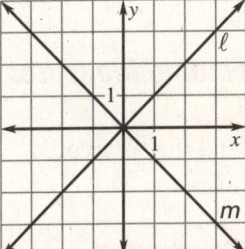

LESSON 11.1

NAME _____ DATE _____

Practice with Examples
For use with pages 661–668

GOAL Find the measures of interior and exterior angles of polygons

VOCABULARY

Theorem 11.1 Polygon Interior Angles Theorem
The sum of the measures of the interior angles of a convex n-gon is $(n - 2) \cdot 180°$.

Corollary to Theorem 11.1
The measure of each interior angle of a regular n-gon is

$$\frac{1}{n} \cdot (n - 2) \cdot 180°, \text{ or } \frac{(n - 2) \cdot 180°}{n}.$$

Theorem 11.2 Polygon Exterior Angles Theorem
The sum of the measures of the exterior angles of a convex polygon, one angle at each vertex, is 360°.

Corollary to Theorem 11.2
The measure of each exterior angle of a regular n-gon is

$$\frac{1}{n} \cdot 360°, \text{ or } \frac{360°}{n}.$$

EXAMPLE 1 Finding Measures of Interior Angles of Polygons

Find the value of x.

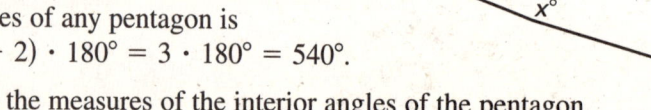

SOLUTION

The sum of the measure of the interior angles of any pentagon is
$(5 - 2) \cdot 180° = 3 \cdot 180° = 540°$.

Add the measures of the interior angles of the pentagon.

$64° + 115° + 96° + 90° + x° = 540°$ The sum is 540°.

$365 + x = 540$ Simplify.

$x = 175$ Subtract 365 from each side.

LESSON 11.1 CONTINUED

Practice with Examples

For use with pages 661–668

Exercises for Example 1

In Exercises 1–3, find the value of x.

1.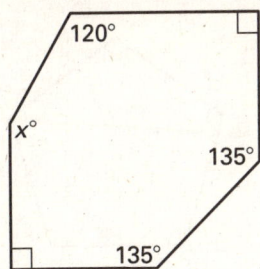
2.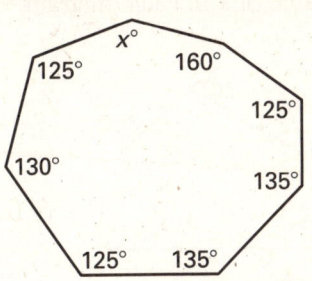
3.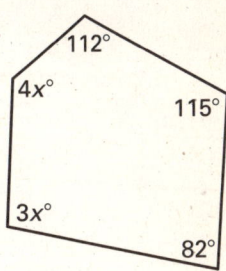

EXAMPLE 2 Finding the Number of Sides of a Polygon

The measure of each interior angle of a regular polygon is 144°. How many sides does the polygon have?

SOLUTION

$\frac{1}{n} \cdot (n-2) \cdot 180° = 144°$ Corollary to Theorem 11.1

$(n-2) \cdot 180 = 144n$ Multiply each side by n.

$180n - 360 = 144n$ Distributive property

$n = 10$ Solve for n.

Exercise for Example 2

4. The measure of each interior angle of a regular n-gon is 156°. Find the value of n.

LESSON 11.1 CONTINUED

Practice with Examples
For use with pages 661–668

EXAMPLE 3 *Finding the Measure of an Exterior Angle*

Find the value of x in each diagram.

a.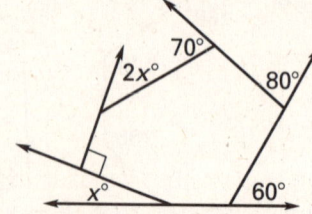

b.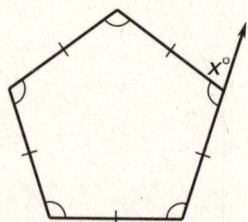

SOLUTION

a. $x° + 90° + 2x° + 70° + 80° + 60° = 360°$ Theorem 11.2
   $3x = 60$ Combine like terms.
   $x = 20$ Divide each side by 3.

b. $x° = \frac{1}{5} \cdot 360°$ Use $n = 5$ in the Corollary to Theorem 11.2.
   $x = 72$ Simplify.

Exercises for Example 3

Find the value of x.

5.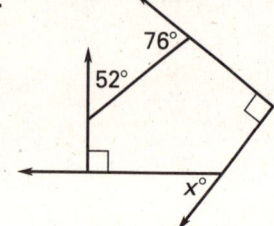

6.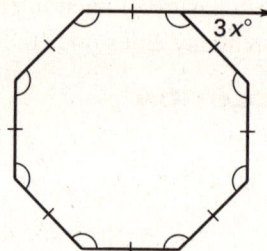

204 **Geometry**
Practice Workbook with Examples

LESSON 11.2

Practice with Examples
For use with pages 669–675

GOAL Find the area of an equilateral triangle and a regular polygon

VOCABULARY

The **center of a regular polygon** is the center of its circumscribed circle.

The **radius of a regular polygon** is the radius of its circumscribed circle.

The distance from the center to any side of a regular polygon is called the **apothem of the polygon**.

A **central angle of a regular polygon** is an angle whose vertex is the center and whose sides contain two consecutive vertices of the polygon.

Theorem 11.3 Area of an Equilateral Triangle The area of an equilateral triangle is one fourth the square of the length of the side times $\sqrt{3}$.
$A = \frac{1}{4}\sqrt{3}s^2$

Theorem 11.4 Area of a Regular Polygon The area of a regular n-gon with side length s is half the product of the apothem a and the perimeter P, so $A = \frac{1}{2}aP$, or $A = \frac{1}{2}a \cdot ns$.

EXAMPLE 1 Finding the Area of an Equilateral Triangle

Find the area of an equilateral triangle with 4 foot sides.

SOLUTION

Use $s = 4$ in the formula of Theorem 11.3.

$$A = \frac{1}{4}\sqrt{3}s^2 = \frac{1}{4}\sqrt{3}(4^2)$$

$$= \frac{1}{4}\sqrt{3}(16) = \frac{1}{4}(16)\sqrt{3} = 4\sqrt{3} \text{ square feet}$$

Using a calculator, the area is about 6.9 square feet.

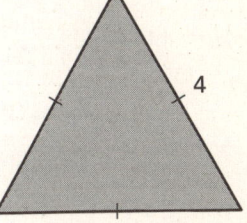

LESSON 11.2 CONTINUED

Practice with Examples
For use with pages 669–675

Exercises for Example 1
Find the area of the triangle.

1.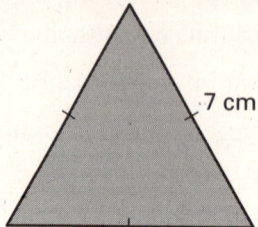

2.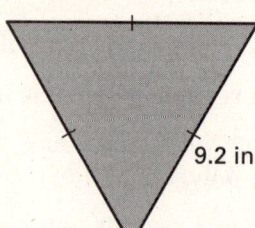

3.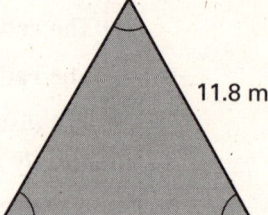

EXAMPLE 2 Finding the Area of a Regular Polygon

A regular octagon is inscribed in a circle with radius 2 units. Find the area of the octagon.

SOLUTION

To apply the formula for the area of a regular octagon, you must find its apothem and perimeter.

The measure of central $\angle ABC$ is $\frac{1}{8} \cdot 360° = 45°$.

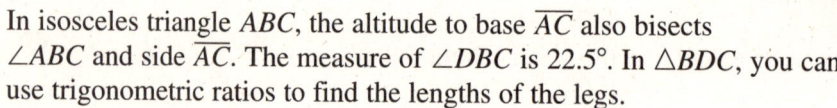

In isosceles triangle ABC, the altitude to base $\overline{AC}$ also bisects $\angle ABC$ and side $\overline{AC}$. The measure of $\angle DBC$ is 22.5°. In $\triangle BDC$, you can use trigonometric ratios to find the lengths of the legs.

$$\cos 22.5° = \frac{BD}{BC} = \frac{BD}{2} \text{ and } \sin 22.5° = \frac{DC}{BC} = \frac{DC}{2}$$

So, the octagon has an apothem of $a = BD = 2 \cdot \cos 22.5°$ and perimeter of $P = 8(AC) = 8(2 \cdot DC) = 8(2 \cdot 2 \cdot \sin 22.5°) = 32 \cdot \sin 22.5°$. The area of the octagon is

$$A = \frac{1}{2}aP = \frac{1}{2}(2 \cdot \cos 22.5°)(32 \cdot \sin 22.5°) \approx 11.3 \text{ square units.}$$

LESSON 11.2 CONTINUED

Practice with Examples
For use with pages 669–675

Exercise for Example 2

4. Find the area of a regular pentagon inscribed in a circle with radius 3 units.

EXAMPLE 3 — Finding the Perimeter and Area of a Regular Polygon

Find the perimeter and area of a regular hexagon with side length of 4 cm and radius 4 cm.

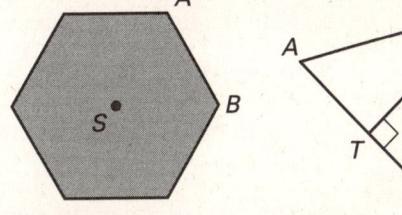

SOLUTION

A hexagon has 6 sides. So, the perimeter is $P = 6(4) = 24$ cm.

To determine the apothem, consider the triangle SBT.

$BT = \frac{1}{2}(BA) = \frac{1}{2}(4) = 2$ cm.

Use the Pythagorean Theorem to find the apothem ST.

$a = \sqrt{4^2 - 2^2} = 2\sqrt{3}$ cm.

So, the area of the hexagon is $A = \frac{1}{2}aP = \frac{1}{2}(2\sqrt{3})(24) = 24\sqrt{3}$ cm².

Exercise for Example 3

Find the perimeter and area of the regular polygon described.

5. Regular octagon with side length 9.18 feet and radius 12 feet.

LESSON 11.3

NAME _____ DATE _____

Practice with Examples

For use with pages 677–682

GOAL Compare perimeters and areas of similar figures and use perimeters and areas of similar figures to solve real-life problems

VOCABULARY

Theorem 11.5   Areas of Similar Polygons
If two polygons are similar with the lengths of corresponding sides in the ratio of $a:b$, then the ratio of their areas is $a^2:b^2$.

EXAMPLE 1   Finding Ratios of Similar Polygons

Hexagons ABCDEF and LMNPQR are similar.

a. Find the ratio of the perimeters of the hexagons.

b. Find the ratio of the areas of the hexagons.

SOLUTION

The ratio of the lengths of corresponding sides in the hexagons is $\frac{3}{7}$, or 3:7.

a. The ratio of the perimeters is also 3:7. So, the perimeter of hexagon ABCDEF is $\frac{3}{7}$ of the perimeter of hexagon LMNPQR.

b. Using Theorem 11.5, the ratio of the areas is $3^2:7^2$, or 9:49. So, the area of hexagon ABCDEF is $\frac{9}{49}$ times the area of hexagon LMNPQR.

Exercises for Example 1

In Exercises 1–4, the polygons are similar. Find the ratio of their perimeters and of their areas.

1. △ABC ~ △DEF

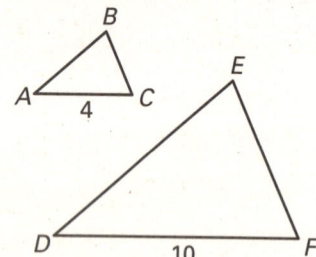

2. ABCD ~ GHEF

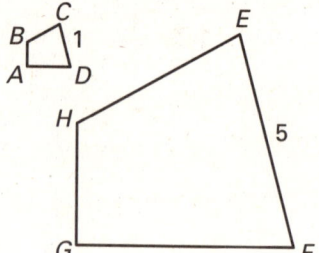

LESSON 11.3 CONTINUED

Practice with Examples

For use with pages 677–682

3. JKLMN ~ PQRST

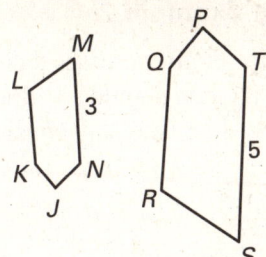

4. △LKJ ~ △XYZ

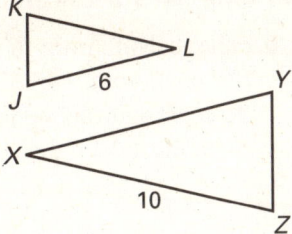

EXAMPLE 2 Using Areas of Similar Figures

A train set is designed to be $\frac{1}{12}$ actual size.

a. A billboard for the train set measures 5.4 inches by 3.2 inches. What would the dimensions of the billboard be in real life?

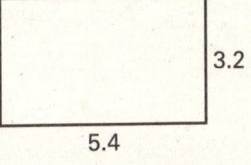

b. If it took Joe 6 minutes to paint the billboard for the train set, what is a rough approximation of how long it would take to paint the real-life billboard?

c. If the perimeter of the track for Joe's train were 18 feet, how long would the track be in real life?

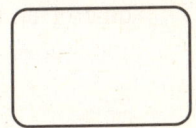

P = 18

SOLUTION

a. The two billboards would be similar figures because they are both rectangles and the scaling of the train set models are proportionate to their real-life counterparts. So, the dimensions of the billboard in real life would be 5.4(12) = 64.8 inches by 3.2(12) = 38.4 inches, or 5.4 feet by 3.2 feet.

b. The ratio of the areas of the rectangles is $1^2:12^2$, or 1:144. Because the amount of time it takes to paint the billboard should be a function of its area, the larger billboard should take about 144 times as long to paint. This would be 864 minutes, or 14.4 hours.

c. The ratio of perimeters of the tracks is the same as the ratio of the individual lengths, or 1:12. So, the perimeter of the real-life track would be 18(12) = 216 feet.

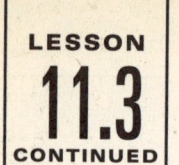

NAME _____ DATE _____

Practice with Examples
For use with pages 677–682

Exercises for Example 2

In Exercises 5–9, refer to the situation in Example 2.

5. The real-life train that Joe modeled his train set after crosses a circular pond with radius 20.5 feet. How big of a circle should Joe draw to represent this pond in his model train set, if he is to stay consistent with the 1:12 ratio?

6. If it takes Joe's train about 7 seconds to traverse the pond in his model set, what is a good approximation of how long it would take the real-life train to cross the real pond?

7. Joe purchased some blue suede material to use as the surface of his pond. The amount of suede used to cover the pond cost about $13.10. What is a good estimate of the cost of enough suede to cover the real-life pond?

8. The area of the real pond is known to be approximately 1320.25 square feet. Use the scale ratio to approximate the area of Joe's model pond, in square feet.

9. Joe created a model of a cow that measured 11 inches long. Is this a good, proportionate representation of a real-life cow? Explain.

LESSON 11.4

NAME _____ DATE _____

Practice with Examples

For use with pages 683–689

GOAL Find the circumference of a circle and the length of a circular arc

VOCABULARY

The **circumference** of a circle is the distance around the circle.

An **arc length** is a portion of the circumference of a circle.

Theorem 11.6 Circumference of a Circle The circumference C of a circle is $C = \pi d$ or $C = 2\pi r$, where d is the diameter of the circle and r is the radius of the circle.

Arc Length Corollary In a circle, the ratio of the length of a given arc to the circumference is equal to the ratio of the measure of the arc to 360°.

$$\frac{\text{Arc length of } \widehat{AB}}{2\pi r} = \frac{m\widehat{AB}}{360°}, \text{ or Arc length of } \widehat{AB} = \frac{m\widehat{AB}}{360°} \cdot 2\pi r$$

EXAMPLE 1 Using Circumferences

a. Find the circumference of a circle with radius 10.5 inches.

b. Find the radius of a circle with circumference 25 feet.

SOLUTION

a. $C = 2\pi r$

$C = 2 \cdot \pi \cdot (10.5)$

$C = 21\pi$

$C \approx 65.97$ inches

b. $C = 2\pi r$

$25 = 2\pi r$

$\dfrac{25}{2\pi} = r$

$r \approx 3.98$ feet

Exercises for Example 1

In Exercises 1–4, find the indicated measure.

1. Find the circumference of a circle with radius 17 centimeters.

2. Find the circumference of a circle with diameter 14 inches.

3. Find the radius of a circle with circumference 14 yards.

4. Find the diameter of a circle with circumference 12 feet.

Geometry
Practice Workbook with Examples

LESSON 11.4 CONTINUED

NAME _____ DATE _____

Practice with Examples
For use with pages 683–689

EXAMPLE 2 Finding Arc Lengths

Find the length of each arc.

a.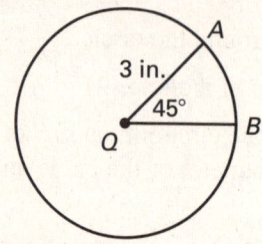

b.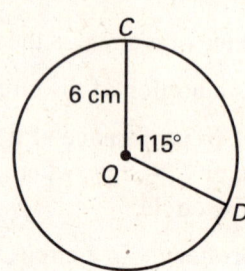

SOLUTION

a. Arc length of $\widehat{AB} = \dfrac{45°}{360°} \cdot 2\pi(3) \approx 2.36$ inches

b. Arc length of $\widehat{CD} = \dfrac{115°}{360°} \cdot 2\pi(6) \approx 12.04$ centimeters

Exercises for Example 2

In Exercises 5–7, find the length of each arc.

5.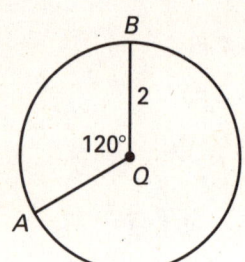

6.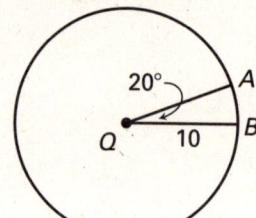

7.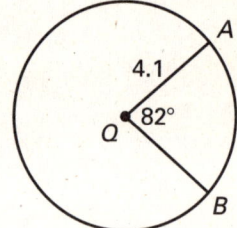

LESSON 11.4 CONTINUED

NAME _____ **DATE** _____

Practice with Examples
For use with pages 683–689

EXAMPLE 3 Using Arc Lengths

Find the circumference of the circle.

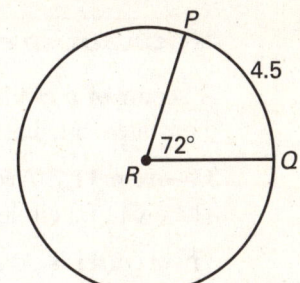

SOLUTION

$$\frac{\text{Arc length of } \widehat{PQ}}{2\pi r} = \frac{m\widehat{PQ}}{360°}$$

Now substitute 4.5 for the arc length of $\widehat{PQ}$, 72° for $m\widehat{PQ}$, and C for $2\pi r$.

So, $\frac{4.5}{C} = \frac{72°}{360°}$, or $\frac{4.5}{C} = 0.2$. So, $C = \frac{4.5}{0.2} = 22.5$.

Exercises for Example 3

Find the indicated measure.

8. Circumference

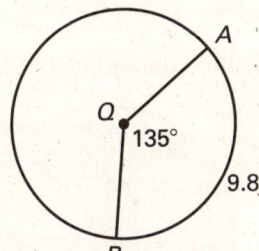

9. Radius

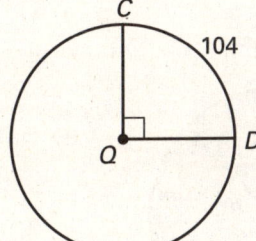

10. $m\widehat{PQ}$

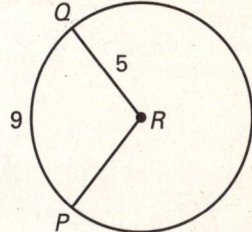

LESSON 11.5

NAME _____ DATE _____

Practice with Examples
For use with pages 691–698

GOAL Find the area of a circle and a sector of a circle and use areas of circles and sectors to solve problems

---

VOCABULARY

A **sector of a circle** is the region bounded by two radii of the circle and their intercepted arc.

Theorem 11.7 Area of a Circle
The area of a circle is π times the square of the radius, or $A = \pi r^2$.

Theorem 11.8 Area of a Sector
The ratio of the area A of a sector of a circle to the area of the circle is equal to the ratio of the measure of the intercepted arc to 360°.

$$\frac{A}{\pi r^2} = \frac{m\widehat{AB}}{360°}, \text{ or } A = \frac{m\widehat{AB}}{360°} \cdot \pi r^2$$

EXAMPLE 1 *Using the Area of a Circle*

a. Find the area of $\odot C$.

b. Find the radius of $\odot P$.

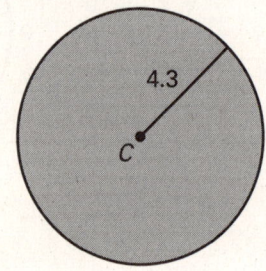
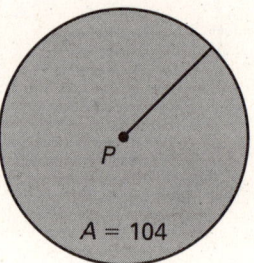

SOLUTION

a. Use $r = 4.3$ in the area formula.

$A = \pi r^2$

$A = \pi \cdot 4.3^2$

$A \approx 58.09$

So, the area is about 58.09 square units.

b. $A = \pi r^2$

$104 = \pi r^2$

$\frac{104}{\pi} = r^2$

$33.10 \approx r^2$

$r \approx 5.75$

LESSON 11.5 CONTINUED

NAME _____ DATE _____

Practice with Examples
For use with pages 691–698

Exercises for Example 1
Find the indicated measure.

1. Area

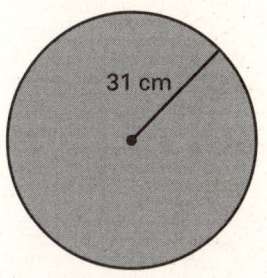

2. Diameter

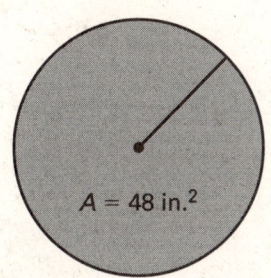

3. Radius

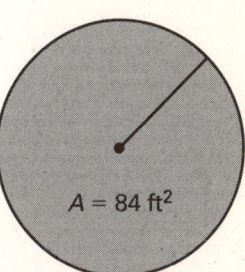

EXAMPLE 2 ### Finding the Area of a Sector

Find the area of the sector shown at the right.

SOLUTION

Sector *CPD* intercepts an arc whose measure is 135°. The radius is 6 centimeters.

$A = \dfrac{m\widehat{CD}}{360°} \cdot \pi r^2$ Write the formula for the area of a sector.

$A = \dfrac{135°}{360°} \cdot \pi \cdot 6^2$ Substitute known values.

$A \approx 42.4$ Use a calculator.

Geometry
Practice Workbook with Examples   **215**

LESSON 11.5 CONTINUED

NAME _____ DATE _____

Practice with Examples
For use with pages 691–698

Exercises for Example 2

In Exercises 4–6, find the area of the shaded region.

4.
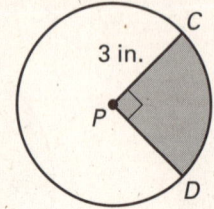

5.
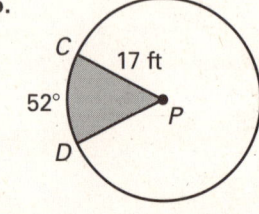

6.
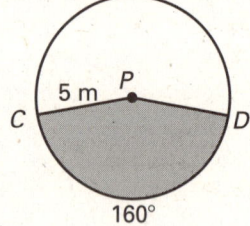

LESSON 11.6

Practice with Examples
For use with pages 699–705

GOAL Find a geometric probability

VOCABULARY

A **probability** is a number from 0 to 1 that represents the chance that an event will occur.

Geometric probability is a probability that involves a geometric measure such as length or area.

Probability and Length Let $\overline{AB}$ be a segment that contains the segment $\overline{CD}$. If a point K on $\overline{AB}$ is chosen at random, then the probability that it is on $\overline{CD}$ is as follows:

$$P(\text{Point } K \text{ is on } \overline{CD}) = \frac{CD}{AB} = \frac{\text{Length of } \overline{CD}}{\text{Length of } \overline{AB}}$$

Probability and Area Let J be a region that contains region M. If a point K in J is chosen at random, then the probability that it is in region M is as follows:

$$P(\text{Point } K \text{ is in region } M) = \frac{\text{Area of } M}{\text{Area of } J}$$

EXAMPLE 1 Finding a Geometric Probability

Find the probability that a point chosen at random on $\overline{AB}$ is on $\overline{CD}$.

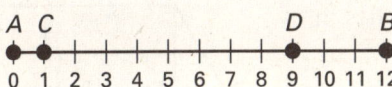

SOLUTION

$$P(\text{Point is on } \overline{CD}) = \frac{\text{Length of } \overline{CD}}{\text{Length of } \overline{AB}} = \frac{8}{12} = \frac{2}{3}$$

The probability can be written as $\frac{2}{3}$, or approximately 0.667, or 66.7%.

LESSON 11.6 CONTINUED

Practice with Examples
For use with pages 699–705

Exercises for Example 1

In Exercises 1–4, find the probability that a point A, selected randomly on $\overline{AB}$, is on the given segment.

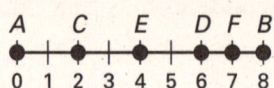

1. $\overline{CD}$ 2. $\overline{EF}$

3. $\overline{CF}$ 4. $\overline{CE}$

EXAMPLE 2 Using Areas to Find a Geometric Probability

Find the probability that a point chosen at random in parallelogram $ABCD$ lies in the shaded region.

SOLUTION

Find the ratio of the area of the shaded square to the area of the parallelogram.

$P(\text{point is in shaded region}) = \dfrac{\text{Area of shaded region}}{\text{Area of parallelogram}}$

$= \dfrac{s^2}{bh} = \dfrac{5^2}{8(5)} = \dfrac{25}{40} = \dfrac{5}{8} = 0.625$

The probability that a point chosen at random in parallelogram $ABCD$ lies in the square is 0.625.

LESSON 11.6 CONTINUED

NAME _____ DATE _____

Practice with Examples

For use with pages 699–705

Exercises for Example 2

Find the probability that a point chosen at random in the figure lies in the shaded region.

5.

6.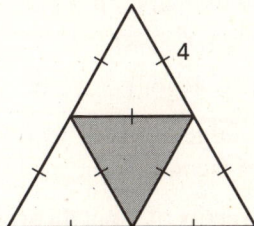

7.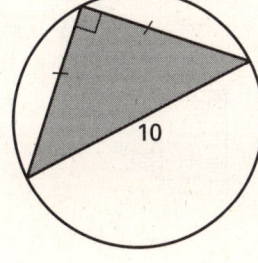

LESSON 12.1

Practice with Examples

For use with pages 719–726

GOAL Use properties of polyhedra and use Euler's Theorem

VOCABULARY

A **polyhedron** is a solid that is bounded by polygons that enclose a single region of space.

The polygons that a polyhedron is bounded by are called **faces**.

An **edge** of a polyhedron is a line segment formed by the intersection of two faces.

A **vertex** of a polyhedron is a point where three or more edges meet.

A polyhedron is **regular** if all of its faces are congruent regular polygons.

Theorem 12.1 Euler's Theorem
The number of faces (F), vertices (V), and edges (E) of a polyhedron are related by the formula $F + V = E + 2$.

EXAMPLE 1 Identifying Polyhedra

Determine whether each solid is a polyhedron. Explain your reasoning.

a.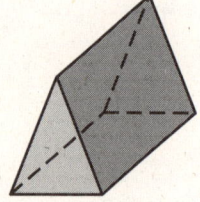

b.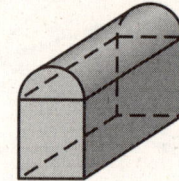

SOLUTION

a. This is a polyhedron. All of its faces are polygons (2 triangles and 3 rectangles), which form a solid enclosing a single region of space.

b. This is not a polyhedron. Some of its faces are not polygons.

LESSON 12.1 CONTINUED

NAME _____ DATE _____

Practice with Examples

For use with pages 719–726

Exercises for Example 1

Determine whether each solid is a polyhedron. Explain your reasoning.

1.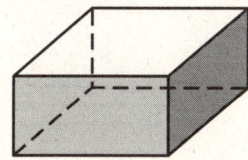
2.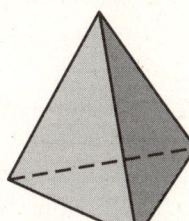
3.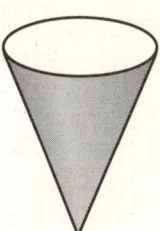

EXAMPLE 2 Analyzing Solids

For each polyhedron, count the number of faces, vertices, and edges.

a.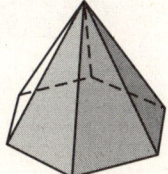
b.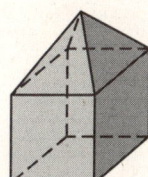

SOLUTION

a. The polyhedron has 7 faces, 7 vertices, and 12 edges.

b. The polyhedron has 9 faces, 9 vertices, and 16 edges.

Exercises for Example 2

Count the number of faces, vertices, and edges.

4.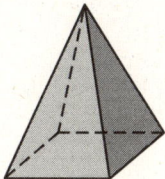
5.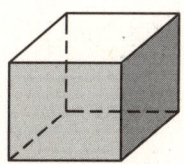
6.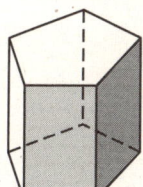

LESSON 12.1 CONTINUED

Practice with Examples

For use with pages 719–726

EXAMPLE 3 Using Euler's Theorem

Calculate the number of vertices of the solid, given that it has 10 faces, all triangles.

SOLUTION

The 10 triangles alone would have $10(3) = 30$ edges. Because each side in the solid is shared by two of these triangles, the total number of edges in the solid is half of this, or 15. Now use Euler's Theorem to find the number of vertices.

$F + V = E + 2$ Write Euler's Theorem.

$10 + V = 15 + 2$ Substitute.

$V = 7$ Solve for V.

Exercise for Example 3

7. Calculate the number of vertices of the solid, given that it has 7 faces; 2 pentagons and 5 triangles.

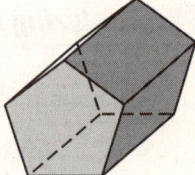

LESSON 12.2

Practice with Examples
For use with pages 728–734

GOAL Find the surface area of a prism and find the surface area of a cylinder

VOCABULARY

A **prism** is a polyhedron with two congruent faces, called **bases**, that lie in parallel planes. The other faces are parallelograms formed by connecting the corresponding vertices of the bases and are called **lateral faces**.

In a **right prism**, each lateral edge is perpendicular to both bases.

Prisms that have lateral edges that are not perpendicular to the bases are **oblique prisms**.

The **surface area of a polyhedron** is the sum of the areas of its faces.

The **lateral area of a polyhedron** is the sum of the areas of its lateral faces.

The two-dimensional representation of all of a prism's faces is called a **net**.

A **cylinder** is a solid with congruent circular bases that lie in parallel planes.

A cylinder is called a **right cylinder** if the segment joining the centers of the bases is perpendicular to the bases.

The **lateral area of a cylinder** is the area of its curved surface and is equal to the product of the circumference and the height.

The entire **surface area of a cylinder** is equal to the sum of the lateral area and the areas of the two bases.

Theorem 12.2 Surface Area of a Right Prism
The surface area S of a right prism can be found using the formula $S = 2B + Ph$, where B is the area of a base, P is the perimeter of a base, and h is the height.

Theorem 12.3 Surface Area of a Right Cylinder
The surface area S of a right cylinder is

$$S = 2B + Ch = 2\pi r^2 + 2\pi rh,$$

where B is the area of a base, C is the circumference of a base, r is the radius of a base, and h is the height.

LESSON 12.2 CONTINUED

Practice with Examples
For use with pages 728–734

EXAMPLE 1 Finding the Surface Area of a Prism Using Theorem 12.2

Find the surface area of the right prism.

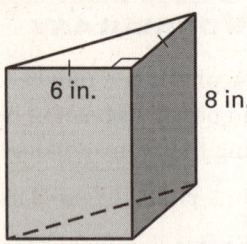

SOLUTION

Each base of the prism is a triangle with base and height of 6 inches. Using the formula for the area of a triangle, the area of each base is $B = \frac{1}{2}(6)(6) = 18$ square inches.

To find the perimeter of each base, you need to find the length of the third side of the triangle. Because the triangle is an isosceles right triangle, it is a 45°-45°-90° triangle. So, the hypotenuse is $6\sqrt{2}$. The perimeter of each base is $P = 6 + 6 + 6\sqrt{2} = 12 + 6\sqrt{2}$. So, the surface area is

$$S = 2B + Ph = 2(18) + (12 + 6\sqrt{2})(8) \approx 199.9 \text{ square inches.}$$

Exercises for Example 1

Find the surface area of the right prism.

1.

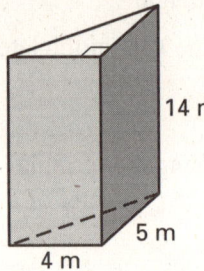

2.

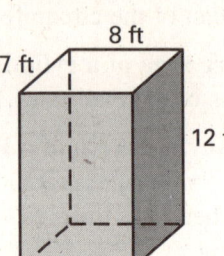

3.

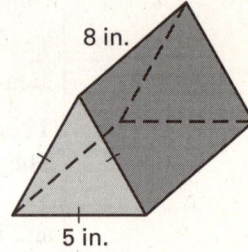

LESSON 12.2 CONTINUED

Practice with Examples

For use with pages 728–734

EXAMPLE 2 Finding the Surface Area of a Cylinder

Find the surface area of the right cylinder.

SOLUTION

The cylinder has an 8 cm radius and a 4 cm height.

$S = 2\pi r^2 + 2\pi rh$ Formula for surface area of a right cylinder

$ = 2\pi(8)^2 + 2\pi(8)(4)$ Substitute.

$ = 128\pi + 64\pi$ Simplify.

$ = 192\pi$ Add.

$ \approx 603.2$ Use a calculator.

The surface area is about 603.2 square centimeters.

Exercises for Example 2

Find the surface area of the right cylinder.

4.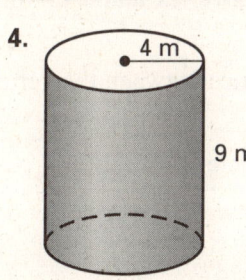

5.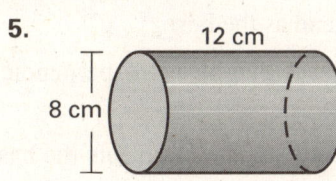

6.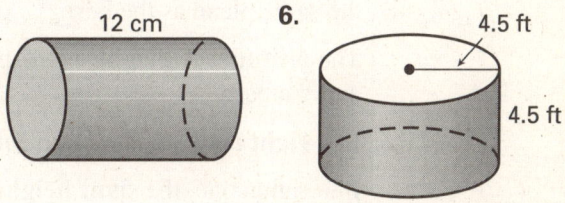

LESSON 12.3

Practice with Examples
For use with pages 735–742

GOAL Find the surface area of a pyramid and find the surface area of a cone

VOCABULARY

A **pyramid** is a polyhedron in which the base is a polygon and the lateral faces are triangles with a common vertex.

The intersection of two lateral faces of a pyramid is called a **lateral edge**.

The intersection of the base of a pyramid and a lateral face is called a **base edge**.

The **altitude**, or **height**, of a pyramid is the perpendicular distance between the base and the vertex.

A **regular pyramid** has a regular polygon for a base and its height meets the base at its center.

The **slant height** of a regular pyramid is the altitude of any lateral face.

A **circular cone**, or **cone**, has a circular base and a vertex that is not in the same plane as the base.

The **altitude**, or **height**, is the perpendicular distance between the vertex and the base.

In a **right cone**, the height meets the base at its center.

In a right cone, the **slant height** is the distance between the vertex and a point on the base edge.

The **lateral surface** of a cone consists of all segments that connect the vertex with points on the base edge.

Theorem 12.4 Surface Area of a Regular Pyramid
The surface area S of a regular pyramid is $S = B + \frac{1}{2}P\ell$, where B is the area of the base, P is the perimeter of the base, and ℓ is the slant height.

Theorem 12.5 Surface Area of a Right Cone
The surface area S of a right cone is $S = \pi r^2 + \pi r \ell$, where r is the radius of the base and ℓ is the slant height.

LESSON 12.3 CONTINUED

Practice with Examples
For use with pages 735–742

EXAMPLE 1 — Finding the Surface Area of a Pyramid

Find the surface area of the regular pyramid shown.

SOLUTION

To find the surface area of the regular pyramid, start by finding the area of the base.

Because the base is a square, the area is s^2. So, the base is 5^2, or 25 square feet.

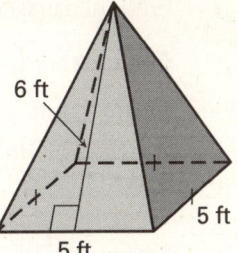

Now you can find the surface area, using 25 for the area of the base, B.

$S = B + \frac{1}{2}P\ell$ Write formula.

$= 25 + \frac{1}{2}(4 \cdot 5)(6)$ Substitute.

$= 85$ Simplify.

So, the surface area is 85 square feet.

Exercises for Example 1

Find the surface area of the regular pyramid.

1.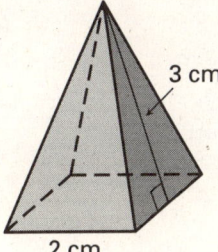

2.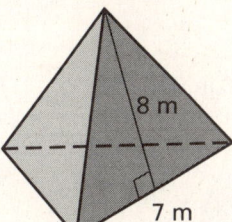

3.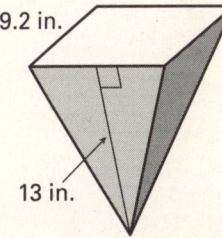

LESSON 12.3 CONTINUED

NAME _____ DATE _____

Practice with Examples
For use with pages 735–742

EXAMPLE 2 **Finding the Surface Area of a Cone**

Find the surface area of the right cone shown.

SOLUTION

With a radius of 3 meters and a slant height of 7 meters given, use the formula to find the surface area.

$S = \pi r^2 + \pi r \ell$ Write formula.

$= \pi(3)^2 + \pi(3)(7)$ Substitute.

$= 9\pi + 21\pi$ Simplify.

$= 30\pi$ Simplify.

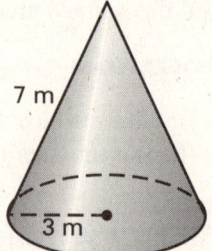

So, the surface area of the cone is 30π square meters, or about 94.2 square meters.

Exercises for Example 2

Find the surface area of the right cone.

4.
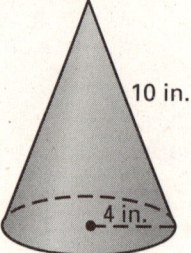

5.
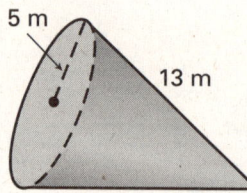

6.
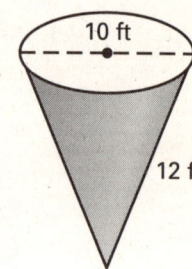

LESSON 12.4

Practice with Examples
For use with pages 743–749

GOAL Use volume postulates and find the volume of prisms and cylinders

VOCABULARY

The **volume of a solid** is the number of cubic units contained in its interior. Volume is measured in cubic units.

Postulate 27 Volume of a Cube The volume of a cube is the cube of the length of its side, or $V = s^3$.

Postulate 28 Volume Congruence Postulate If two polyhedra are congruent, then they have the same volume.

Postulate 29 Volume Addition Postulate The volume of a solid is the sum of the volumes of all its nonoverlapping parts.

Theorem 12.6 Cavalieri's Principle If two solids have the same height and the same cross-sectional area at every level, then they have the same volume.

Theorem 12.7 Volume of a Prism The volume V of a prism is $V = Bh$, where B is the area of a base and h is the height.

Theorem 12.8 Volume of a Cylinder The volume V of a cylinder is $V = Bh = \pi r^2 h$, where B is the area of a base, h is the height, and r is the radius of a base.

EXAMPLE 1 Finding Volumes

Find the volume of the right cylinder and the right prism.

a.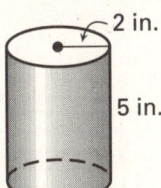

b.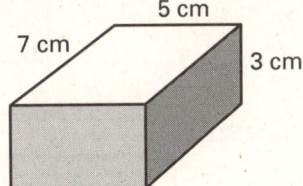

SOLUTION

a. The area B of the base is $\pi \cdot 2^2$, or 4π in.^{2}. Use $h = 5$ to find the volume.

$$V = Bh = 4\pi(5) = 20\pi \approx 62.83 \text{ in.}^3$$

b. The area B of the base is $(7)(5)$, or 35 cm^2. Use $h = 3$ to find the volume.

$$V = Bh = (35)(3) = 105 \text{ cm}^3$$

LESSON 12.4 CONTINUED

Practice with Examples
For use with pages 743–749

Exercises for Example 1
Find the volume of the right prism or the right cylinder.

1.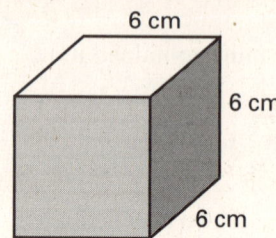
2.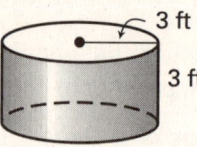
3.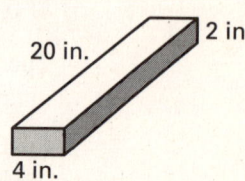

EXAMPLE 2 Using Volumes

Use the measurements given to solve for x.

$V = 60$ cm^3

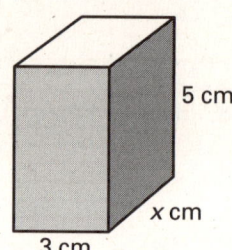

SOLUTION

The area of the base is $3x$ square centimeters.

$V = Bh$	Formula for volume of a right prism
$60 = (3x)(5)$	Substitute.
$60 = 15x$	Rewrite.
$\dfrac{60}{15} = x$	Divide each side by 15.
$4 = x$	Simplify.

LESSON 12.4 CONTINUED

Practice with Examples

For use with pages 743–749

Exercises for Example 2

Use the measurements given to solve for x.

4. $V = 283 \text{ m}^3$

5. $V = 64 \text{ in.}^3$

6. $V = 300 \text{ ft}^3$

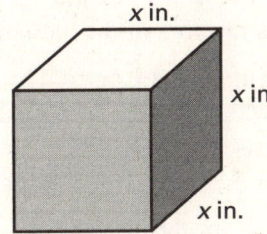
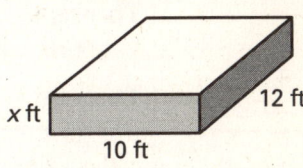

LESSON 12.5

NAME _____ DATE _____

Practice with Examples

For use with pages 752–758

GOAL Find the volume of pyramids and cones

VOCABULARY

Theorem 12.9 Volume of a Pyramid The volume V of a pyramid is $V = \frac{1}{3}Bh$, where B is the area of the base and h is the height.

Theorem 12.10 Volume of a Cone The volume V of a cone is $V = \frac{1}{3}Bh = \frac{1}{3}\pi r^2 h$, where B is the area of the base, h is the height, and r is the radius of the base.

EXAMPLE 1 Finding the Volume of a Pyramid

Find the volume of the pyramid with the square base shown to the right.

SOLUTION

The area B of the base of the pyramid is the area of the square. Using the formula for the area of a square, s^2, $B = 11^2$, or 121 square centimeters. Using $h = 21$, you can find the volume.

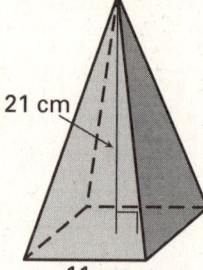

$V = \dfrac{1}{3}Bh$ Formula for volume of pyramid

$= \dfrac{1}{3}(121)(21)$ Substitute.

$= 847$ Simplify.

So, the volume of the pyramid is 847 cubic centimeters.

Exercises for Example 1

In Exercises 1–3, find the volume of the pyramid.

1.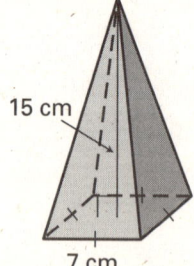

2.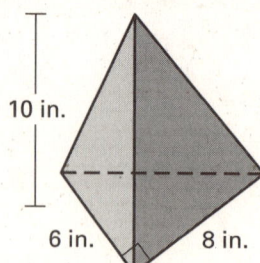

3.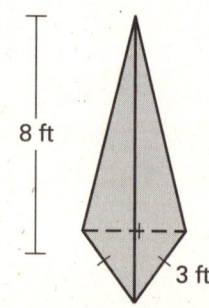

LESSON 12.5 CONTINUED

NAME _____ DATE _____

Practice with Examples

For use with pages 752–758

EXAMPLE 2 Finding the Volume of a Cone

Find the volume of the cone.

SOLUTION

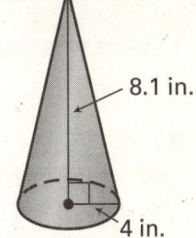

$V = \frac{1}{3}Bh = \frac{1}{3}(\pi r^2)h$ Formula for volume of cone

$= \frac{1}{3}(\pi \cdot 4^2)(8.1)$ Substitute.

$= 43.2\pi$ Simplify.

So, the volume of the cone is 43.2π in.^{3}, or about 135.7 in.^{3}

Exercises for Example 2

Find the volume of the cone.

4.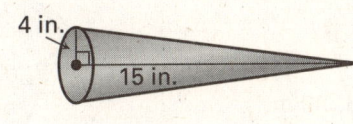

5.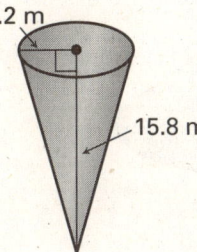

6.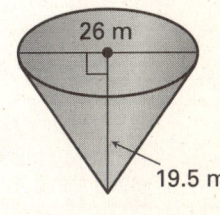

LESSON 12.5 CONTINUED

Practice with Examples
For use with pages 752–758

EXAMPLE 3 Using the Volume of a Cone

Use the given measurements to solve for x.

SOLUTION

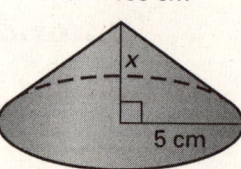
$V = 105$ cm³
5 cm

$V = \frac{1}{3}\pi r^2 h$ Formula for volume of cone

$105 = \frac{1}{3}\pi \cdot 5^2 \cdot x$ Substitute.

$4 \approx x$ Simplify and solve for x.

The height of the cone is about 4 centimeters.

Exercises for Example 3

In Exercises 7–9, find the value of x.

7. $V = 182$ in.³
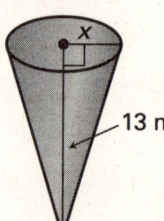
6 in.

8. $V = 215$ m³
13 m

9. $V = 56.5$ m³
6 m

LESSON 12.6

NAME _____ DATE _____

Practice with Examples

For use with pages 759–765

GOAL Find the surface area of a sphere and find the volume of a sphere

VOCABULARY

A **sphere** is the locus of points in space that are a given distance from a point called the **center of the sphere**.

A **radius of a sphere** is a segment from the center to a point on the sphere.

A **chord of a sphere** is a segment whose endpoints are on the sphere.

A **diameter of a sphere** is a chord that contains the center.

If a plane that intersects a sphere contains the center of the sphere, the intersection is a **great circle** of the sphere.

A great circle of a sphere separates the sphere into two congruent halves called **hemispheres**.

Theorem 12.11 Surface Area of a Sphere The surface area S of a sphere with radius r is $S = 4\pi r^2$.

Theorem 12.12 Volume of a Sphere The volume V of a sphere with radius r is $V = \frac{4}{3}\pi r^3$.

EXAMPLE 1 Finding the Surface Area of a Sphere

Find the surface area of the sphere.

SOLUTION

$S = 4\pi r^2$ Formula for surface area of sphere

$= 4\pi(10)^2$ Substitute.

$= 400\pi$ Simplify.

So, the surface area of the sphere is 400π square feet, or about 1256.6 square feet.

LESSON 12.6 CONTINUED

Practice with Examples

For use with pages 759–765

Exercises for Example 1

Find the surface area of the sphere.

1.
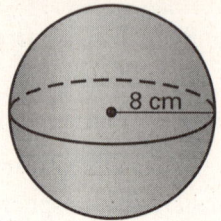
8 cm

2.
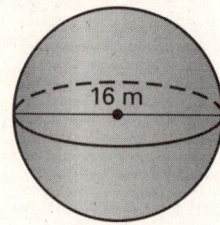
16 m

3.
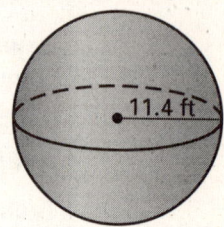
11.4 ft

EXAMPLE 2 Using a Great Circle

The circumference of a great circle of a sphere is 25 inches. Find the surface area of the sphere.

SOLUTION

Begin by finding the radius of the sphere.

$C = 2\pi r$ Formula for circumference of a circle

$25 = 2\pi r$ Substitute.

$4 \approx r$ Divide each side by 2π.

Using a radius of 4 cm, the surface area is $S = 4\pi r^2 = 4\pi(4)^2 = 64\pi$ in.²

So, the surface area of the sphere is 64π in.², or about 201.1 in.²

Exercises for Example 2

Find the surface area of the sphere.

4.
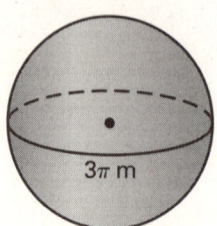
3π m

5.
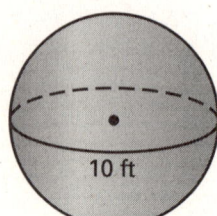
10 ft

6.
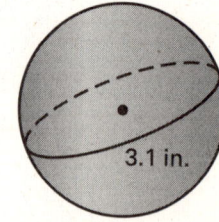
3.1 in.

Geometry
Practice Workbook with Examples

LESSON 12.6 CONTINUED

NAME _____ **DATE** _____

Practice with Examples

For use with pages 759–765

EXAMPLE 3 Finding the Volume of a Sphere

Find the volume of the sphere.

SOLUTION

$V = \dfrac{4}{3}\pi r^3$ Formula for volume of sphere

$ = \dfrac{4}{3}\pi (3.5)^3$ Substitute.

$ \approx 179.6$ Simplify.

So, the volume of the sphere is about 179.6 cubic feet.

Exercises for Example 3

Find the volume of the sphere.

7. 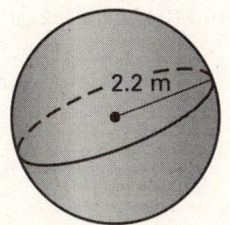 2.2 m

8. 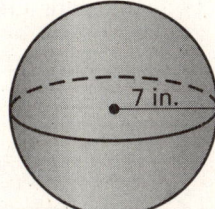 7 in.

9. 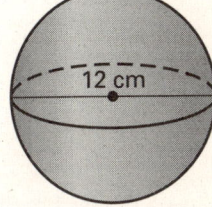 12 cm

LESSON 12.7

Practice with Examples
For use with pages 766–772

GOAL Find and use the scale factor of similar solids and use similar solids to solve problems

VOCABULARY

Two solids with equal ratios of corresponding linear measures, such as heights or radii, are called **similar solids**.

The common ratio of linear measures for a pair of similar solids is called the **scale factor** of one solid to the other solid.

Theorem 12.13 Similar Solids Theorem If two similar solids have a scale factor of $a:b$, then corresponding areas have a ratio of $a^2:b^2$, and corresponding volumes have a ratio of $a^3:b^3$.

EXAMPLE 1 Identifying Similar Solids

Decide whether the two solids are similar. If so, find the scale factor.

a.
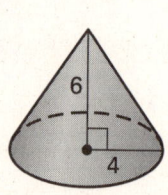

b.
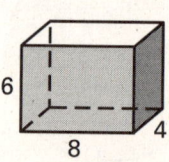 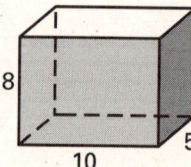

SOLUTION

a. The solids are similar because the ratios of corresponding linear measures are equal, as shown.

$$\text{radii: } \frac{4}{6} = \frac{2}{3} \qquad \text{heights: } \frac{6}{9} = \frac{2}{3}$$

The solids have a scale factor of 2:3.

b. The solids are not similar because the ratios of corresponding linear measures are not equal, as shown.

$$\text{lengths: } \frac{4}{5} \qquad \text{widths: } \frac{8}{10} = \frac{4}{5} \qquad \text{heights: } \frac{6}{8} = \frac{3}{4}$$

LESSON 12.7 CONTINUED

Practice with Examples
For use with pages 766–772

Exercises for Example 1

Decide whether the two solids are similar. If so, find the scale factor.

1.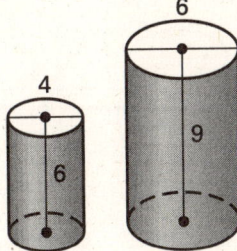
2.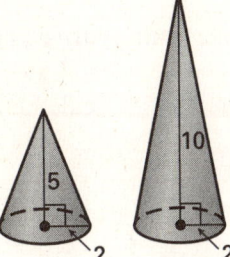
3.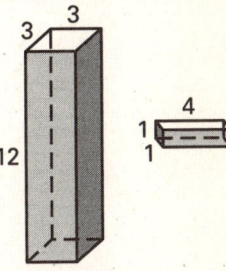

EXAMPLE 2 Using the Scale Factor of Similar Solids

The spheres are similar with a scale factor of 1:4. Find the surface area and volume of sphere B given that the surface area of sphere A is 144π square inches and the volume of sphere A is 288π cubic inches.

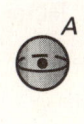
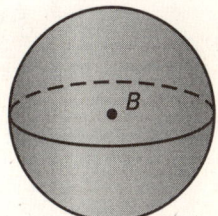

SOLUTION

Begin by using Theorem 12.13 to set up two proportions.

$$\frac{\text{Surface area of } A}{\text{Surface area of } B} = \frac{a^2}{b^2} \qquad \frac{\text{Volume of } A}{\text{Volume of } B} = \frac{a^3}{b^3}$$

$$\frac{144\pi}{\text{Surface area of } B} = \frac{1^2}{4^2} \qquad \frac{288\pi}{\text{Volume of } B} = \frac{1^3}{4^3}$$

Surface area of $B = 2304\pi$ \qquad Volume of $B = 18{,}432\pi$

So, the surface area of sphere B is 2304π square inches and the volume of sphere B is $18{,}432\pi$ cubic inches.

LESSON 12.7 CONTINUED

Practice with Examples
For use with pages 766–772

Exercises for Example 2

The solid described is similar to a larger solid with the given scale factor. Find the surface area *S* and volume *V* of the larger solid.

4. A right cylinder with a surface area of 48π square centimeters and a volume of 45π cubic centimeters; scale factor 2:3

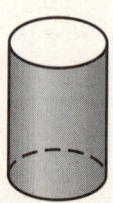

5. A right prism with a surface area of 82 square feet and a volume of 42 cubic feet; scale factor 1:2

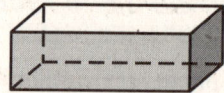

ANSWERS

Chapter 1

Lesson 1.1

1.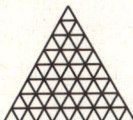

2. The second number is 2 times the first, the third number is 3 times the second, and so on; 120

3. *Sample*: The first number is 1 squared then minus 1, the second number is 2 squared then minus 1, the third number is 3 squared then minus 1, and so on; 35

4. $a^2 - b^2$

5. *Sample*: $(4 + 5)^2 = 81 \neq 41 = 4^2 + 5^2$

Lesson 1.2

1.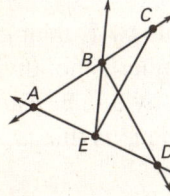

2, 4, 6, 7. Sample answers are given. **2.** A, B, D
3. $\overrightarrow{BA}, \overrightarrow{BC}$ **4.** $\overrightarrow{BD}$
5. A, B, C **6.** $\overrightarrow{BA}, \overrightarrow{BD}$
7. $\overleftrightarrow{AB}, \overleftrightarrow{BC}$

8.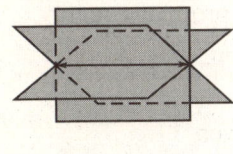

9.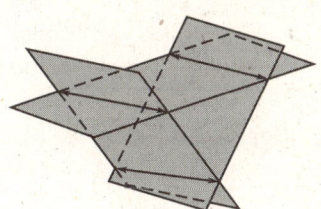

Lesson 1.3

1. a. 10 b. 5 c. $\overline{AB}$ and $\overline{BC}$ are congruent.
2. a. 4 b. 4 c. 5 d. 5 3. $7\sqrt{5}$
4. $\sqrt{61}$ 5. $\sqrt{1261}$ 6. $4\sqrt{10}$ 7. $\sqrt{89}$
8. $2\sqrt{a^2 + b^2}$

Lesson 1.4

1. $\angle ABC, \angle CBA, \angle B$; vertex B, sides $\overrightarrow{BA}$ and $\overrightarrow{BC}$ 2. $\angle DEF, \angle FED, \angle E$; vertex E, sides $\overrightarrow{EF}$ and $\overrightarrow{ED}$ 3. 35°

4. a. acute b. obtuse
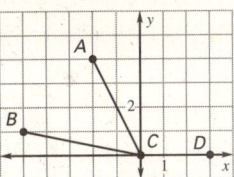 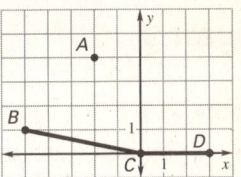

c. obtuse
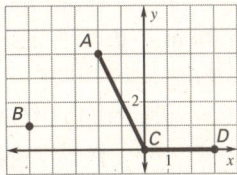

5. a. straight b. acute
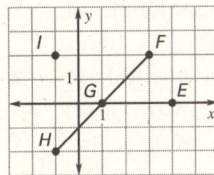 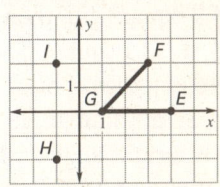

c. obtuse d. right
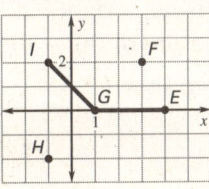 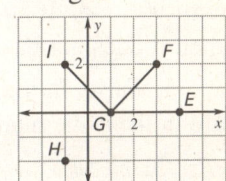

e. right
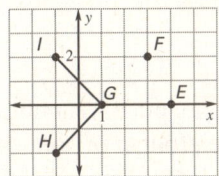

Lesson 1.5

1. (2.5, 1.5) 2. (−0.5, 7.5) 3. (−2.5, 1.5)
4. (1, −3) 5. (0, 24) 6. (7, 3) 7. 5 8. 10

Lesson 1.6

1. a. no b. no c. yes d. no 2. a. yes
b. no c. yes d. no 3. $x = 15$,
$m\angle AEB = 90°, m\angle BEC = 90°, m\angle CED = 90°,$
$m\angle AED = 90°$ 4. $y = 40, m\angle FJG = 150°,$
$m\angle GJH = 30°, m\angle HJI = 150°, m\angle FJI = 30°$
5. 9° 6. 153°

Chapter 1 continued

Lesson 1.7
1. 75.36 units, 452.16 square units   2. 48 units, 84 square units   3. 21.71 ft, 24.53 ft²

Chapter 2

Lesson 2.1
1. **a.** If you will wash the dishes, then I will dry them.   **b.** If you will not wash the dishes, then I will not dry them.   **c.** If I will dry the dishes, then you will wash them.   **d.** If I will not dry the dishes, then you will not wash them.
2. **a.** If a square has side length of 3 cm, then it has an area of 9 square cm.   **b.** If a square does not have side length of 3 cm, then it does not have an area of 9 square cm.   **c.** If a square has an area of 9 square cm, then it has side length of 3 cm.   **d.** If a square does not have an area of 9 square cm, then it does not have a side length of 3 cm.   3. **a.** If an angle has a measure of 90°, then it is a right angle.   **b.** If an angle does not have a measure of 90°, then it is not a right angle.   **c.** If an angle is a right angle, then it has a measure of 90°.   **d.** If an angle is not a right angle, then it does not have a measure of 90°.
4. true   5. False. In the diagram shown, $\overleftrightarrow{AB}$ and C are in the same plane.

Lesson 2.2
1. conditional statement: If a number is a perfect square, then it is the product of some number times itself; converse: If a number is the product of some number and itself, then it is a perfect square.   2. conditional statement: If two angles are complementary, then the sum of their measures is 90°; converse: If the sum of the measures of two angles is 90°, then the angles are complementary.   3. conditional statement: If a real number is rational, then it can be written in the form $\frac{p}{q}$, where p and q are integers and $q \neq 0$; converse: If a real number can be written in the form $\frac{p}{q}$, where p and q are integers and $q \neq 0$, then it is a rational number.

4. **a.** The statement is a biconditional statement because it contains the words "if and only if."   **b.** false   5. **a.** The statement is a biconditional statement because it contains the words "if and only if."   **b.** true   6. If angles are vertical angles, then the sides of the two angles form two pairs of opposite rays; True; The sides of two angles form two pairs of opposite rays if and only if the angles are vertical angles.   7. If a is 0 or b is 0, then the product ab is 0; True; The product of ab is 0 if and only if either a is 0 or b is 0.

Lesson 2.3
1. **a.** If two lines in a plane are parallel, then they do not intersect.   **b.** If two lines in a plane do not intersect, then they are parallel.   **c.** true
2. **a.** If you are in North America, then you are in the United States.   **b.** If you are in the United States, then you are in North America.   **c.** false   3. **a.** If two angles are not supplementary, then the measures of the angles do not sum to 180°.   **b.** If the measures of two angles do not sum to 180°, then the angles are not supplementary.
4. **a.** If a number is not divisible by 5, then it is not divisible by 10.   **b.** If a number is not divisible by 10, then it is not divisible by 5.   5. False. We do not know whether or not Pete is thirsty.

Lesson 2.4
1. $2x + 3 = 7x$ Given
$3 = 5x$ Subtraction prop. of equality
$\frac{3}{5} = x$ Division prop. of equality

2.
$4 + 2(3x + 5) = 11 - x$ Given
$4 + 6x + 10 = 11 - x$ Distributive prop.
$6x + 14 = 11 - x$ Simplify.
$6x = -3 - x$ Subtr. prop. of equal.
$7x = -3$ Add. prop. of equal.
$x = -\frac{3}{7}$ Div. prop. of equality

3. $6x - 2 = -4(x - 1)$ Given
$6x - 2 = -4x + 4$ Distributive prop.
$6x = -4x + 6$ Add. prop. of equal.
$10x = 6$ Add. prop. of equal.
$x = \frac{3}{5}$ Div. prop. of equal.

Chapter 2 continued

4.
$\frac{1}{5}x + 4 = 2x + \frac{3}{5}$ Given
$\frac{1}{5}x + \frac{17}{5} = 2x$ Subt. prop. of eq.
$\frac{17}{5} = \frac{9}{5}x$ Subt. prop. of eq.
$\frac{17}{9} = x$ Mult. prop. of eq.

5.
MN = PQ   Given
MP = MN + NP   Segment Addition Postulate
MP = PQ + NP   Substitution prop. of equality
NQ = PQ + NP   Segment Addition Postulate
MP = NQ   Substitution prop. of equality

6.
AB = DE   Given
AD = AB + BD   Segment Addition Postulate
AD = DE + BD   Substitution prop. of equality
BE = DE + BD   Segment Addition Postulate
AD = BE   Substitution prop. of equality

7.
$m\angle AQB = m\angle CQD$ Given
$m\angle AQC = m\angle AQB + m\angle BQC$ Angle Addition Postulate
$m\angle AQC = m\angle CQD + m\angle BQC$ Substitution prop. of equality
$m\angle BQD = m\angle CQD + m\angle BQC$ Angle Addition Postulate
$m\angle AQC = m\angle BQD$ Substitution prop. of equality

8.
$m\angle RPS = m\angle TPV$ Given
$m\angle TPV = m\angle SPT$ Given
$m\angle RPS = m\angle SPT$ Transitive prop. of equality
$m\angle RPV = m\angle RPS + m\angle SPT + m\angle TPV$ Angle Addition Postulate
$m\angle RPV = m\angle RPS + m\angle RPS + m\angle RPS$ Substitution prop. of equality
$m\angle RPV = 3(m\angle RPS)$ Distributive property

Lesson 2.5

1.

Statements	Reasons
1. $AD = AB = 12$	1. Given, Transitive
2. $\overline{AD} \cong \overline{AB}$	2. Def. of congruent segments
3. $\overline{AD} \cong \overline{CD}, \overline{CD} \cong \overline{BC}$	3. Given
4. $\overline{AB} \cong \overline{BC}$	4. Transitive prop. of congruence

2, 3. Check steps. **2.** 3   **3.** 3

Lesson 2.6

1–6. Check explanations.   **1.** 90°   **2.** 90°   **3.** 31°
4. 125°   **5.** 180°   **6.** 52°   **7.** 90°; 36°
8. 132°; 48°

Chapter 3

Lesson 3.1

1. skew   **2.** parallel   **3.** parallel
4. perpendicular   **5.** $\overleftrightarrow{AD}, \overleftrightarrow{EH}, \overleftrightarrow{DC}, \overleftrightarrow{HG}$
6. ABE   **7.** $\overleftrightarrow{FG}$ **8.** $\overrightarrow{AE}, \overrightarrow{DH}, \overrightarrow{AD}, \overrightarrow{AB}, \overrightarrow{DC}$
9. alternate interior   **10.** alternate exterior
11. consecutive interior   **12.** corresponding
13. consecutive interior   **14.** alternate interior

Lesson 3.2

1. Statements
1. $\overrightarrow{BA} \perp \overrightarrow{BC}$
2. $\angle ABC$ is a right $\angle$.
3. $m\angle ABC = 90°$
4. $m\angle 3 + m\angle 4 = m\angle ABC$
5. $m\angle 3 + m\angle 4 = 90°$
6. $\angle 3$ and $\angle 4$ are complementary.

Reasons
1. Given
2. Definition of $\perp$ lines
3. Definition of right angle
4. Angle Addition Postulate
5. Substitution Property of Equality
6. Definition of complementary angles

Geometry
Practice Workbook with Examples

Chapter 3 *continued*

2. Since it is given that ∠1 and ∠2 are a linear pair, ∠1 and ∠2 are also supplementary by the Linear Pair Post., so $m\angle 1 + m\angle 2 = 180°$. Since j is perpendicular to k, $m\angle 1 = 90°$. By the substitution and subtraction properties, you can conclude that $m\angle 2$ is also 90°, or a right angle. This same argument can be used with the linear pair of ∠1 and ∠4 to show that ∠4 is a right angle, too. Lastly, since ∠1 and ∠3 are vertical angles and vertical angles are congruent, ∠3 has the same measure as ∠1, or 90°, a right angle. So, ∠1, ∠2, ∠3, and ∠4 are all right angles.

3. 45 **4.** 90 **5.** 30

Lesson 3.3

1. 113° **2.** 67° **3.** 113° **4.** 67° **5.** 113°
6. 67° **7.** 113° **8.** 68 **9.** 25 **10.** 12
11. 10 **12.** 5 **13.** 5 **14.** 12 **15.** 12

Lesson 3.4

1. The measure of ∠1 is 120° because the 60° angle and ∠1 form a linear pair. Thus ℓ and m are parallel by Theorem 3.10.

2. The measure of ∠2 is 135° by the Vertical Angles Theorem. Consecutive interior angles are supplementary ($m\angle 2 + 45° = 180°$). Thus n and o are parallel by Theorem 3.9. **3.** 120 **4.** 30 **5.** 15

Lesson 3.5

1. Alternate exterior angles are congruent.

2. k is perpendicular to both of the given parallel lines. Since ℓ is perpendicular to at least one of the same lines, k and ℓ are parallel. **3.** ∠1 and the $x°$ angle form a linear pair, so the measure of ∠1 is $180 - x$, which equals the corresponding angle formed by $70 + (110 - x)$.
Since corresponding angles are congruent, k is parallel to ℓ. **4.** a and b **5.** e and f
6. j and ℓ, m and n

Lesson 3.6

1. 3 **2.** 5 **3.** -3 **4.** $\frac{1}{3}$ **5.** $\frac{3}{4}$ **6.** 0
7. $-\frac{1}{12}, \frac{1}{12}$, not parallel **8.** $\frac{1}{2}, \frac{1}{2}$, parallel
9. $y = x - 7$ **10.** $y = -x - 3$
11. $y = \frac{2}{3}x + \frac{8}{3}$

Lesson 3.7

1. yes **2.** no **3.** $y = \frac{7}{4}x + \frac{19}{2}$
4. $y = -\frac{2}{5}x - \frac{3}{2}$ **5.** $y = 5x + 7$

Chapter 4

Lesson 4.1

1. right scalene **2.** equiangular equilateral
3. obtuse isosceles **4.** 30 **5.** 10

Lesson 4.2

1. a. 3 **b.** 5 **2. a.** 10 **b.** 2 **3.** 65 **4.** 2

Lesson 4.3

1. The marks on the diagram show that the three sides are congruent. By the SSS Congruence Postulate, the two triangles are congruent.

2. Because the top and bottom lines are parallel, alternate interior angles are congruent. For the two triangles, one common side is shared, and it is congruent to itself. By SAS, the two triangles are congruent. **3.** $AB = DE = \sqrt{5}$ so, $\overline{AB} \cong \overline{DE}$; $BC = EF = \sqrt{13}$ so, $\overline{BC} \cong \overline{EF}$; $CA = FD = 2\sqrt{5}$ so, $\overline{CA} \cong \overline{FD}$; By SSS, $\triangle ABC \cong \triangle DEF$.

Lesson 4.4

1.

Statements	Reasons
1. $\overline{MC} \cong \overline{AC}$	1. Given
2. ∠NMC and ∠BAC are right angles.	2. Given
3. ∠NMC ≅ ∠BAC	3. Right angles are ≅
4. ∠MCN ≅ ∠ACB	4. Vertical angles are ≅
5. △NMC ≅ △BAC	5. ASA Congruence Post.

Chapter 4 continued

2.

Statements	Reasons
1. $\overline{AE} \cong \overline{DE}$	1. Given
2. $\angle A \cong \angle D$	2. Given
3. $\angle AEB \cong \angle DEC$	3. Vertical angles are $\cong$
4. $\triangle BAE \cong \triangle CDE$	4. ASA Congruence Post.

3.

Statements	Reasons
1. $\angle G \cong \angle B$	1. Given
2. $\overline{CB} \parallel \overline{GA}$	2. Given
3. $\angle BCA \cong \angle GAC$	3. Alternate Int. $\angle$s Thm.
4. $\overline{AC} \cong \overline{AC}$	4. Reflexive prop. of $\cong$
5. $\triangle GCA \cong \triangle BAC$	5. AAS Congruence Thm.

4.

Statements	Reasons
1. $\angle LMO \cong \angle JNO$	1. Given
2. $\angle MOL \cong \angle NOJ$	2. Vertical angles are $\cong$
3. $\angle L \cong \angle J$	3. Third Angles Theorem
4. $\angle OMN \cong \angle ONM$	4. Given
5. $\overline{MN} \cong \overline{MN}$	5. Reflexive prop. of $\cong$
6. $\triangle MJN \cong \triangle NLM$	6. AAS Congruence Thm.

Lesson 4.5

1.

Statements	Reasons
1. $\overline{RT} \cong \overline{AS}$, $\overline{RS} \cong \overline{AT}$	1. Given
2. $\overline{ST} \cong \overline{ST}$	2. Reflexive Prop. of $\cong$
3. $\triangle TSA \cong \triangle STR$	3. SSS Congruence Post.
4. $\angle TSA \cong \angle STR$	4. Corresponding parts of $\cong$ triangles are $\cong$

2. You are given $\angle 1 \cong \angle 3$, $\angle 4 \cong \angle 5$, and $\overline{ES} \cong \overline{DT}$. By ASA, $\triangle HES \cong \triangle HDT$. Because corresponding parts of congruent triangles are congruent, $\overline{HE} \cong \overline{HD}$.

3. You are given $\overline{PA}$ congruent to $\overline{KA}$ and $\overline{LA}$ congruent to $\overline{NA}$. These are corresponding sides to the larger triangles PAL and KAN. The angles PAL and KAN of these triangles are congruent because they are vertical angles. Thus, triangles PAL and KAN are congruent by SAS. Angles KNA and PLA are congruent because they are corresponding parts of congruent triangles. Now, angles NAY and LAX are congruent because they are vertical angles. By ASA, triangles NAY and LAX are congruent. Finally, because corresponding parts of $\cong$ $\triangle$s are $\cong$, $\overline{AX}$ is congruent to $\overline{AY}$.

4. Because $\overline{DC}$ is congruent to $\overline{BA}$, angles CDL and ABM are congruent, and $\overline{DB}$ is congruent to itself, the larger triangles BAD and DCB are congruent by SAS. Because corresponding parts of $\cong$ $\triangle$ are $\cong$, segments $\overline{DA}$ and $\overline{BC}$ are congruent and angles CBD and ADB are congruent. You are given angles DAL and BCM are congruent. Now, by ASA, triangles CBM and ADL are congruent. Because corresponding parts of $\cong$ $\triangle$s are $\cong$, $\overline{AL}$ is congruent to $\overline{CM}$.

Lesson 4.6

1. You are given that $\overline{BC} \perp \overline{AD}$, so by the definition of perpendicular lines $m\angle ACB = m\angle DCB = 90°$. Thus $\triangle ACB$ and $\triangle DCB$ are right triangles. You are also given that $\overline{AB} \cong \overline{DB}$. $\overline{BC} \cong \overline{BC}$ by the reflexive property of congruence. Thus $\triangle ACB \cong \triangle DCB$ by the Hypotenuse-Leg Congruence Theorem.

2. You are given $m\angle JKL = m\angle MLK = 90°$, so triangles JKL and MLK are right triangles. You are also given that $\overline{JL} \cong \overline{MK}$. $\overline{KL} \cong \overline{KL}$ by the reflexive property of congruence. Thus, $\triangle JKL \cong \triangle MLK$ by the Hypotenuse-Leg Congruence Theorem.

3. $x = 20$ **4.** $x = 4$ **5.** $x = 60$, $y = 4$

Lesson 4.7

1. $2\sqrt{17}$ **2.** $(5, 4)$ **3.** Given the coordinates for both triangles, you can use the distance formula to find the lengths of each side of each triangle. You would then show that corresponding sides are congruent, and by SSS the triangles are congruent.

Chapter 5

Lesson 5.1

1. 6 **2.** 9 **3.** No, If E were on $\overleftrightarrow{AB}$, then E would be equidistant from C and D. E is 14 units from C but 15 units from D. **4.** 6 **5.** 40° **6.** 10 **7.** $\sqrt{73}$ **8.** 16

Lesson 5.2

1. a. 12 **b.** 9 **2. a.** 20 **b.** 12 **c.** 12 **d.** 11 **3.** 3

Chapter 5 continued

Lesson 5.3
1. 6   2. 5   3. 4   4. 26   5. (5, 2)
6. (−5, 6)   7. (2, 6)

Lesson 5.4
1. The midpoints are $D(2, 3)$ and $E(4, 0)$; $BC = 2\sqrt{13} = 2 \cdot ED$; both slopes equal $-\frac{3}{2}$.
2. $DF = 7, CB = 10$ 3. 32
4. $(-6, 11), (-2, -1), (6, -5)$
5. $(-11, 3), (-3, 7), (9, -5)$
6. $(-2, 0), (4, 6), (10, -4)$
7. $(-7, 1), (3, 7), (9, -1)$

Lesson 5.5
1. $m\angle A < m\angle C < m\angle B$; $BC < AB < AC$
2. $DF < DE < EF$; $m\angle E < m\angle F < m\angle D$
3. $m\angle H < m\angle I < m\angle G$; $GI < GH < HI$
4. $KL < JK < JL$; $m\angle J < m\angle L < m\angle K$
5. greater than 3 cm and less than 7 cm
6. greater than 5 in. and less than 19 in.
7. greater than 6 ft and less than 14 ft
8. greater than 1 m and less than 21 m
9. greater than 16 in. and less than 34 in.
10. greater than 7 mi and less than 9 mi

Lesson 5.6
1. <   2. <   3. =   4. =   5. >   6. >   7. 53°

Chapter 6

Lesson 6.1
1. yes   2. No, the one side of the figure is not a segment.   3. yes   4. No, two of the sides only intersect one other side.   5. convex
6. concave   7. concave   8. 90   9. 36

Lesson 6.2
1. $a = 15, b = 135$ 2. $d = 10, e = 90$
3. $x = 24, y = 120$ 4. $a = 5, b = 3, c = 4$
5. $d = 98, e = 98, f = 82$
6. $g = 12, h = 9, i = 16, j = 14$
7. $x = 30, y = 6$ 8. $x = 40, y = 8$
9. $x = 10, y = 9$

Lesson 6.3
1. Slope of $\overline{AB}$ = slope of $\overline{CD}$ = −4; slope of $\overline{AD}$ = slope of $\overline{BC}$ = $\frac{2}{5}$
2. $AB = CD = \sqrt{26}$; $AD = BC = \sqrt{45} = 3\sqrt{5}$
3. Sample answer: Slope of $\overline{AB}$ = slope of $\overline{CD} = \frac{1}{4}$; $AB = CD = \sqrt{17}$ 4. Sample answer: Slope of $\overline{AB}$ = slope of $\overline{CD} = -\frac{1}{6}$; slope of $\overline{AD}$ = slope of $\overline{BC}$ = 9

Lesson 6.4
1. 4 ≅ sides, opposite sides are ∥, opposite ∠s are ≅, consecutive ∠s are supplementary, diagonals bisect each other, diagonals are ⊥, diagonals bisect opposite ∠s.
2. 4 right ∠s, opposite sides are ∥ and ≅, opposite ∠s are ≅, consecutive ∠s are supplementary, diagonals bisect each other, diagonals are ≅.
3. always
4. sometimes
5. always   6. 30   7. 10   8. 29

Lesson 6.5
1. 7   2. 9   3. 118   4. 110   5. 35   6. 5

Lesson 6.6
1. $AB = BC = CD = DA = 3$
2. Slope of $\overline{EF}$ = slope of $\overline{GH} = \frac{1}{4}$; slope of $\overline{FG}$ = slope of $\overline{HE} = -\frac{1}{4}$; so EFGH is a parallelogram; slope of $\overline{EG}$ = 0 and slope of $\overline{FH}$ is undefined, so they are perpendicular.
3. kite   4. isosceles trapezoid   5. rectangle

Lesson 6.7
1. 61 sq units   2. 81 sq units   3. 864 sq units
4. 40 sq units   5. 25 sq units   6. 98 sq units

Chapter 7

Chapter 7

Lesson 7.1
1. reflection in the line $x = -1$
2. $E(0, 0), F(0, 2), G(2, 3), H(2, 0)$
3. A and H, B and G, C and F, D and E
4. no 5. yes 6. $x = 12, y = 4$
7. $x = 40, y = 4$

Lesson 7.2
1.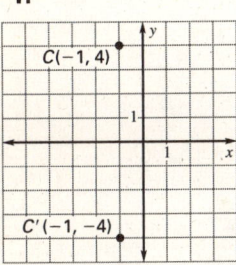
2.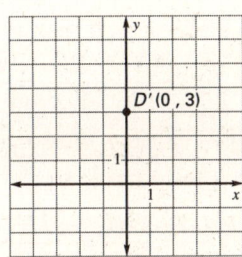
3.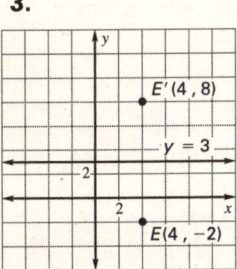
4.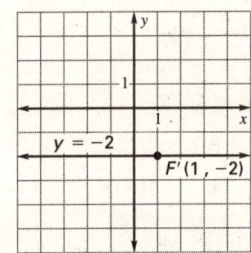
5.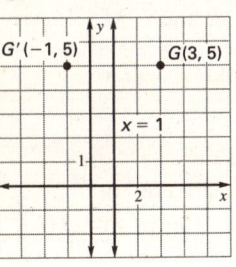
6.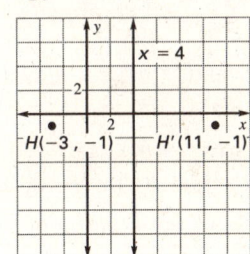
7.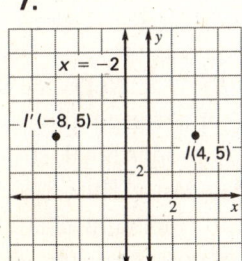
8.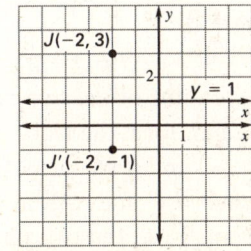

9. two 10. one 11. $(2, 0)$ 12. $(5, 0)$
13. $(1.5, 0)$

Lesson 7.3
1. $A'(2, 2), B'(-1, -2), C'(5, -2), D'(0, -6)$; $(x, y) \longrightarrow (y, -x)$
2. $X'(-2, -5), Y'(-5, -5), Z'(-5, -1), W'(-2, -1); (x, y) \longrightarrow (-x, -y)$ 3. Yes, a rotation of 120° about its center 4. Yes, a rotation of 180° about its center 5. Yes, a rotation of 72° about its center

Lesson 7.4
1. XX', ZZ' 2. 3 cm 3. $X''Y''Z''$ 4. k and m
5.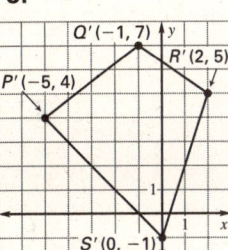
6.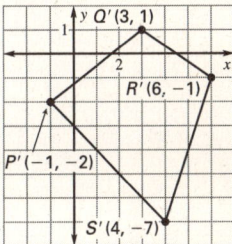
7.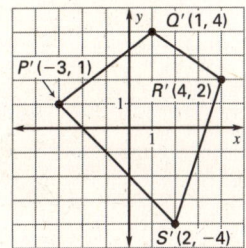
8.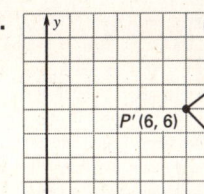

9. $\langle -2, -4 \rangle$ 10. $\langle 3, -3 \rangle$

Lesson 7.5
1. does not affect the image
2. does affect the image
3. does affect the image
4. does not affect the image

Chapter 7 continued

5.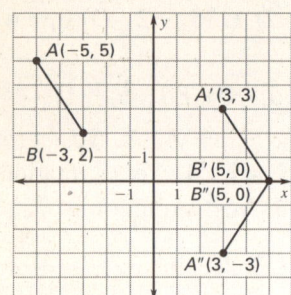

6.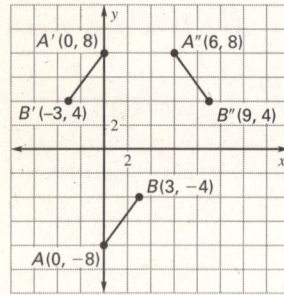

7.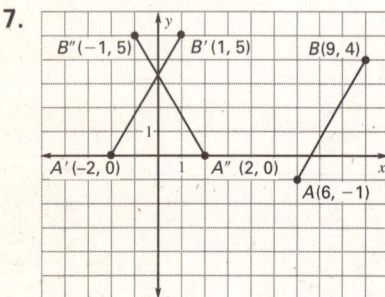

8.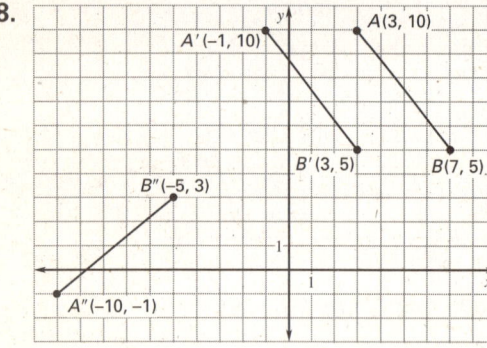

Lesson 7.6

1. TRHVG **2.** T **3.** TRVG **4.** Yes, in the line $y = -\frac{1}{2}$ **5.** Yes, in the line $x = 6$, $x = 13$ **6.** reflection in the line $x = 13$ **7.** horizontal glide reflection

Chapter 8

Lesson 8.1

1. $\frac{1}{8}$ **2.** $\frac{3}{1}$ **3.** $\frac{1}{1}$ **4.** $\frac{2}{3}$ **5.** length = 25 ft, width = 5 ft **6.** 31.5 **7.** 2 **8.** -12 **9.** 6

Lesson 8.2

1. 3.2 **2.** 4 **3.** 9 **4.** 3

Lesson 8.3

1. $\angle A \cong \angle D$, $\angle B \cong \angle E$, $\angle C \cong \angle F$; $\frac{AB}{DE} = \frac{BC}{EF} = \frac{CA}{FD}$

2. $\angle A \cong \angle Z$, $\angle B \cong \angle W$, $\angle D \cong \angle X$, $\angle C \cong \angle Y$; $\frac{AB}{ZW} = \frac{BD}{WX} = \frac{DC}{XY} = \frac{CA}{YZ}$

3. $\angle E \cong \angle M$, $\angle F \cong \angle R$, $\angle G \cong \angle Q$, $\angle H \cong \angle P$, $\angle J \cong \angle N$; $\frac{EF}{MR} = \frac{FG}{RQ} = \frac{GH}{QP} = \frac{HJ}{PN} = \frac{JE}{NM}$

4. Yes, $ABCDE \sim HJKFG$ **5.** no **6.** 11.2
7. 14

Lesson 8.4

1. $\angle A$ and $\angle P$, $\angle C$ and $\angle N$, $\angle B$ and $\angle M$; $\frac{AB}{PM} = \frac{BC}{MN} = \frac{CA}{NP}$; 30 **2.** $\angle XYT$ and $\angle Z$, $\angle XTY$ and $\angle W$, $\angle X$ and $\angle X$; $\frac{XY}{XZ} = \frac{YT}{ZW} = \frac{TX}{WX}$; 4.25 **3.** $\angle K$ and $\angle M$, $\angle KJL$ and $\angle MLJ$, $\angle KLJ$ and $\angle MJL$; $\frac{KL}{MJ} = \frac{LJ}{JL} = \frac{JK}{LM}$; 8.2

4. $\triangle ABD \sim \triangle BCE$ **5.** no; the angles in $\triangle PQN$ are not congruent to the angles in $\triangle MPR$.
6. $\triangle XYT \sim \triangle XWZ$

Lesson 8.5

1. $\triangle ABC \sim \triangle EFD$ **2.** $\triangle MNP \sim \triangle RQS$
3. The lengths of the two pairs of corresponding sides which form the right ∠s are proportional, so the △ are similar by the SAS Similarity Thm.
4. Because $\overline{PQ}$ and $\overline{RN}$ are ∥, corresponding ∠s are congruent. Now, the two pairs of corresponding sides which include the two corresponding ∠s Q and NRM have lengths that are proportional, so the △ are similar by the SAS Similarity Thm.

Chapter 8 continued

5. The two triangles have congruent angles at point S because the angles are vertical angles. The two sides forming these angles are proportional, so the triangles are similar by the SAS Similarity Theorem.

Lesson 8.6
1. 3.5   **2.** $\frac{200}{9}$ **3.** $x \approx 8.9, y \approx 14.1$
4. $a \approx 6.5$ **5.** $x \approx 2.7$ **6.** $x \approx 7.8, y \approx 7.2$

Lesson 8.7
1. $k = \frac{3}{7}$, reduction   **2.** $k = 2$, enlargement
3. $k = \frac{7}{3}$, enlargement   **4.** $k = \frac{2}{5}$, reduction
5. $A'(0, 6), B'(6, 6), C'(4.5, 3)$
6. $A'(-24, -9), B'(-15, -12), C'(-6, -9), D'(-12, -3)$
7. $K'(0.5, 4), L'(1, 2), N'(-3, 0), M'(-5, 0)$
8. $X'(-2.25, -1.5), Y'(6, 3), Z'(3, -3)$

Chapter 9

Lesson 9.1
1. 2.4 units   **2.** 1 unit   **3.** ≈ 5.1 units
4. 6.5 units   **5.** 11.5 units   **6.** 11.4 units

Lesson 9.2
1. 10.6; no   **2.** 20; yes   **3.** 13; yes
4. 11.6 units   **5.** 3 units   **6.** 5.0 units
7. 20.1 square units   **8.** 98.8 square units
9. 9.2 square units

Lesson 9.3
1. yes   **2.** no   **3.** no   **4.** yes; right   **5.** no
6. yes; obtuse   **7.** yes; obtuse

Lesson 9.4
1. $42\sqrt{2} \approx 59.4$ **2.** $9\sqrt{2} \approx 12.7$
3. $x = \frac{13\sqrt{2}}{2} \approx 9.2, y = \frac{13\sqrt{2}}{2} \approx 9.2$
4. $14\sqrt{3} \approx 24.2$ **5.** $7\sqrt{3} \approx 12.1$
6. $x = 5\sqrt{3} \approx 8.7, y = 10\sqrt{3} \approx 17.4$

Lesson 9.5
1. $\sin A = 0.8, \cos A = 0.6, \tan A = 1.3333$
2. $\sin A = 0.7667, \cos A = 0.6333, \tan A = 1.2105$
3. $\sin A = 0.8333, \cos A = 0.5556, \tan A = 1.5$
4. 35.2 feet   **5.** 70.7 meters

Lesson 9.6
1. $x = 14, y \approx 20.7, z = 25, m\angle Y \approx 55.9°, m\angle X \approx 34.1°, m\angle Z = 90°$ **2.** $l = 14, m = 12, n \approx 18.4, m\angle L \approx 49.4°, m\angle M \approx 40.6°, m\angle N = 90°$ **3.** $p = 9.4, q \approx 5.3, r = 10.8, m\angle Q \approx 29.5°, m\angle P \approx 60.5°, m\angle R = 90°$ **4.** $a = 19, b \approx 30.9, c \approx 24.3, m\angle A = 38°, m\angle B = 90°, m\angle C = 52°$
5. $l \approx 5.2, m = 9, n \approx 7.4, m\angle L = 35°, m\angle M = 90°, m\angle N = 55°$ **6.** $p \approx 55.8, q = 41.5, r \approx 37.4, m\angle P = 90°, m\angle Q = 48°, m\angle R = 42°$

Lesson 9.7
1. $\langle -2, 7 \rangle; \approx 7.3$ **2.** $\langle 2, -6 \rangle; \approx 6.3$

3. $\langle -4, 2 \rangle; \approx 4.5$ **4.** $\langle 4, 13 \rangle; \approx 13.6$

5. 9.2, 49.4° north of east
6. 7.1, 45° south of east
7. 7.1, 45° south of west
8. $\langle -3, 13 \rangle$ **9.** $\langle 0, 3 \rangle$ **10.** $\langle 0, 0 \rangle$

Chapter 10

Lesson 10.1
1. chord   **2.** secant   **3.** radius   **4.** diameter
5. tangent   **6.** radius   **7.** chord   **8.** radius

Chapter 10 continued

9. 5 or −5 **10.** 4 **11.** 16

Lesson 10.2
1. **a.** 180° **b.** 90° **c.** 90° **d.** 270°
2. **a.** 110° **b** 200° **c.** 160° **d.** 250°
3. $\sqrt{11} \approx 3.3$ **4.** 3

Lesson 10.3
1. 34 **2.** 23 **3.** 43 **4.** 50 **5.** 25 **6.** 11
7. $x = 40, y = 93$ **8.** $x = 56, y = 20$
9. $x = 30, y \approx 59.3$

Lesson 10.4
1. $x = 78, y = 102$ **2.** $x = 180$
3. $x = 60, y = 240, z = 120$
4. 65 **5.** 70 **6.** 30 **7.** 49 **8.** 70 **9.** 20

Lesson 10.5
1. 3.2 **2.** 4 **3.** 6 **4.** 7.25 **5.** ≈ 3.7 **6.** 2
7. $\sqrt{55} \approx 7.4$ **8.** $8\sqrt{3} \approx 13.9$
9. $\sqrt{\frac{117}{4}} \approx 5.4$

Lesson 10.6
1. $(x - 4)^2 + (y + 1)^2 = 36$
2. $(x + 1)^2 + (y + 5)^2 = 10.24$
3. $(x + 2)^2 + (y - 3)^2 = 16$

4.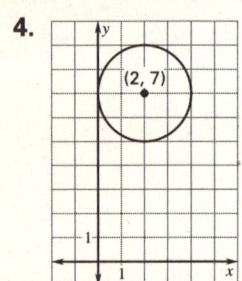
5.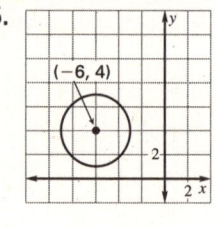
6.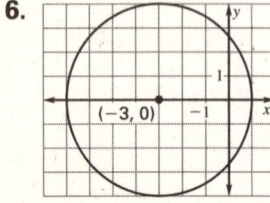
7.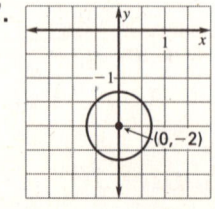

8. no **9.** yes **10.** no

Lesson 10.7
1. The line $y = 5$.
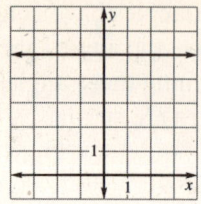

2. The circle $(x + 1)^2 + (y - 2)^2 = 25$ and its interior points.
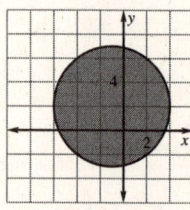

3. The line $y = x$.
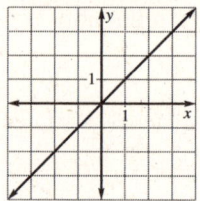

4. $(-1, 1), (3, 1)$
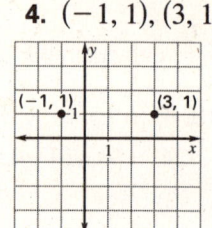

5. segment from $(-4, 0)$ to $(4, 0)$ and segment from $(0, 4)$ to $(0, -4)$
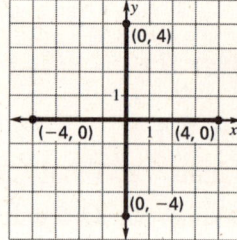

Chapter 11

Lesson 11.1
1. 150 **2.** 145 **3.** 33 **4.** 15 **5.** 52 **6.** 15

Lesson 11.2
1. 21.2 cm² **2.** 36.7 in.² **3.** 60.3 m²
4. 21.4 square units **5.** 73.44 ft, 407.13 ft²

Lesson 11.3
1. 2:5, 4:25 **2.** 1:5, 1:25 **3.** 3:5, 9:25
4. 3:5, 9:25 **5.** 1.7 ft **6.** 84 sec
7. $1886.40 **8.** 9.2 ft² **9.** No, in real life, the cow would be 132 inches long, or 11 feet long. This is not a realistic length for a cow.

Chapter 11 continued

Lesson 11.4
1. 106.81 cm 2. 43.98 in. 3. 2.23 yd
4. 3.82 ft 5. 4.19 6. 3.49 7. 5.87
8. 26.13 9. 66.21 10. 103.1°

Lesson 11.5
1. 3019.07 cm² 2. 7.82 in.² 3. 5.17 ft
4. 7.07 in.² 5. 131.14 ft² 6. 34.91 m²

Lesson 11.6
1. 0.5 2. 0.375 3. 0.625 4. 0.25
5. 0.2146 6. 0.25 7. 0.3183

Chapter 12

Lesson 12.1
1. yes; all of its faces are polygons which form a solid enclosing a single region of space.
2. yes; all of its faces are polygons which form a solid enclosing a single region of space.
3. no; some of its faces are not polygons.
4. 5 faces, 5 vertices, 8 edges
5. 6 faces, 8 vertices, 12 edges
6. 7 faces, 10 vertices, 15 edges 7. 10

Lesson 12.2
1. 235.6 m² 2. 472 ft² 3. 141.7 in.²
4. 326.7 m² 5. 402.1 cm² 6. 254.5 ft²

Lesson 12.3
1. 16 cm² 2. 105.2 m² 3. 323.8 in.²
4. 175.9 in.² 5. 282.7 m² 6. 267 ft²

Lesson 12.4
1. 216 cm³ 2. 84.8 ft³ 3. 160 in.³
4. 3 5. 4 6. 2.5

Lesson 12.5
1. 245 cm³ 2. 80 in.³ 3. 10.4 ft³
4. 251.3 in.³ 5. 1400.4 m³ 6. 3451 m³
7. 4.8 in. 8. 4 m 9. 6 m

Lesson 12.6
1. 804.2 cm² 2. 804.2 m² 3. 1633.1 ft²
4. 28.3 m² 5. 31.8 ft² 6. 3.1 in.²
7. 44.6 m³ 8. 1436.8 in.³ 9. 904.8 cm³

Lesson 12.7
1. yes; 2:3 2. no 3. yes; 1:3
4. 339.3 cm²; 477.1 cm³ 5. 328 ft²; 336 ft³